JOURNAL OF CYBER
SECURITY AND MOBILITY

Volume 1, Nos. 2–3 (April/July 2012)

JOURNAL OF CYBER SECURITY AND MOBILITY

Aim

Journal of Cyber Security and Mobility provides an in-depth and holistic view of security and solutions from practical to theoretical aspects. It covers topics that are equally valuable for practitioners as well as those new in the field.

Scope

The journal covers security issues in cyber space and solutions thereof. As cyber space has moved towards the wireless/mobile world, issues in wireless/mobile communications will also be published. The publication will take a holistic view. Some example topics are: security in mobile networks, security and mobility optimization, cyber security, cloud security, Internet of Things (IoT) and machine-to-machine technologies.

JOURNAL OF CYBER SECURITY AND MOBILITY

Volume 1   Nos. 2–3   April/July 2012

Published, sold and distributed by:
River Publishers
P.O. Box 1657
Algade 42
9000 Aalborg
Denmark

Tel.: +45369953197
www.riverpublishers.com

Journal of Cyber Security and Mobility is published four times a year.
Publication programme, 2012: Volume 1 (4 issues)

ISSN 2245-1439 (Print Version)
ISSN 2245-4578 (Online Version)

Editorial Foreword

Welcome to the second volume of the *Journal of Cyber Security and Mobility*. This issue is sequel to the premier issue of the journal that was published during January 2012 on the eve of second World Wide Security Mobility Conference (WWSMC) at Princeton, NJ. Since security and mobility issues permeate multiple layers and are very much intertwined in all walks of life, it is important to consider the issues affecting these layers, namely physical layer, MAC layer, network layer and application layer and study the areas where security and mobility have greatest performance impact and tradeoff. Recently, security issues for wireless sensor networks, smart grid and cloud computing are also getting a lot of prominence. The second volume of the journal has addressed the security issues applicable to above areas.

The current issue of the journal has seven papers. First three papers provide overall algorithmic solutions in wireless and security space. Papers 4 and 5 focus on security issues at MAC layer and performance. Last two papers are somewhat survey type; paper 6 focuses on smart grid security and the paper 7 focuses on security for cloud computing. First paper by Tange and Andersen highlights secure plain Diffie-Hellman algorithm that solves the man in the middle attack problem. In the second paper Mathur et al. identify fairness constraints, signal to Interference Ratio (SINR) constraint and balance constraint and study their impact on the complexity of the channel assignment problems. Khajuria and Andersen highlight the importance of integration of security services to the low power devices and present a crypto solution for the reconfiguration devices in the third paper. Fourth paper by Pawar et al. models the activities of MAC layer security attacks in the wireless sensor networks (WSN) and shows the performance degradation of WSN under these attacks. In fifth paper, Rohokale et al. highlight the importance of cooperative wireless communications as virtual MIMO technique to combat fading and achieve diversity through user cooperation and undermines the need for physical layer security. Cybersecurity threat is becoming an important concern for every walk of life including smart grids, service providers, defense, financial industries and enterprise markets. Deng and Shukla provide a summary of vulnerabilities and countermeasures that will enable to secure the transmission system from cyber borne threats in the sixth paper. Finally, the seventh paper focuses on security, privacy and trust challenges aspects of cloud computing. This paper by Nayak provides an overview of the challenges of in the area of cloud, illustrates a few real-life case study implementations and subsequent policy considerations. We hope the readers will find these articles timely and interesting.

We would like to thank the reviewers, editorial board members, advisory board members, steering committee members and the staff of River Publishers during the publication process of the second volume of the journal. We hope the readers will continue to find the forthcoming issues of this journal useful and contribute in the area of cybersecurity and mobility.

Editors-in-Chief
Ashutosh Dutta, Niksun
Ruby Lee, Princeton University
Neeli Prasad, Aalborg University

SPDH – A Secure Plain Diffie–Hellman Algorithm

Henrik Tange and Birger Andersen

Center for Wireless Systems and Applications/CTIF-Copenhagen, Copenhagen University College of Engineering, Lautrupvang 15, 2750 Ballerup, Denmark; e-mail: {heta, bia}@ihk.dk

Abstract

Secure communication in a wireless system or end-to-end communication requires setup of a shared secret. This shared secret can be obtained by the use of a public key cryptography system. The most widely used algorithm to obtain a shared secret is the Diffie–Hellman algorithm. However, this algorithm suffers from the Man-in-the-Middle problem; an attacker can perform an eavesdropping attack listen to the communication between participants A and B. Other algorithms as for instance ECMQV (Elliptic Curve Menezes Qo Vanstone) can handle this problem but is far more complex and slower because the algorithm is a three-pass algorithm whereas the Diffie–Hellman algorithm is a simple two-pass algorithm. Using standard cryptographic modules as AES and HMAC the purposed algorithm, Secure Plain Diffie–Hellman Algorithm, solves the Man-in-the-Middle problem and maintain its advantage from the plain Diffie–Hellman algorithm. Also the possibilities of replay attacks are solved by use of a timestamp.

Keywords: secure Diffie–Hellman algorithm, AES, HMAC, Man-in-the-Middle attacks, replay attacks.

Journal of Cyber Security and Mobility, Vol. 1, 143–160.

1 Introduction

Secure communication between two parties over an unsecure channel in a network in general requires confidentiality, data integrity, data origin authentication, non-repudiation and entity reputation.

The use of confidentiality ensures that data are secret for other than the two participants. Data integrity ensures that data have not been altered passing over the unsecure channel. To ensure that the sender of a message is the sender data origin authentication is used. The goal of non-repudiation is to make it able for the receiver to document that the message is sent from the sender. At last, entity reputation convinces the participants of each other's identity.

Furthermore, cryptographic systems can be divided into symmetric key systems and public key cryptography. Symmetric key systems are used for encryption and decryption of a message that should be kept secret. The encryption and decryption is done using a shared secret key. The public key cryptography is used to transport the shared secret key in a safe manner. The public key cryptography must provide a way to solve the key distribution problem also known as a need for private and authenticated key transport over an unsecure channel. This problem is partly solved by the Diffie–Hellman key exchange which makes it possible to obtain privacy in the key exchange. The advantage of the Diffie–Hellman algorithm is that, it is a lightweight two-pass protocol with only a public key transport from participant A to participant B and again from B to A. In the Diffie–Hellman algorithm the public key is used on both sides to calculate the shared secret. The problem is that the Diffie–Hellman algorithm is vulnerable against Man-in-the-Middle attacks. In this paper we show how the Diffie–Hellman algorithm can be protected against Man-in-the-Middle attacks and still function as a lightweight two-pass protocol. We show how the protocol can be used with one or more authentication centers with limited overhead and finally we verify the security of the protocol.

2 Background

Since the first establishment of electronic networks security has been an issue in order to keep information secret between parties. In this background we look into some important algorithms in public key cryptography, symmetric algorithms and message authentication.

2.1 The RSA Algorithm

There are several public key cryptography algorithms. One of the first public key cryptosystems was the RSA (Rivest, Shamir and Adleman) algorithm invented in 1977 [1, pp. 6]. The number theoretical problem behind the RSA algorithm is the integer factorization problem [1, pp. 6–7]. The integer factorization problem is the problem of calculating the plaintext m in the encryption and signature scheme:

$$m^{ed} \equiv m \,(\text{mod}) \, n,$$

where e is an encryption exponent and d is a private key or decryption exponent. m^{ed} is equal to the ciphertext c. Decryption of the ciphertext c is done by using the private key $d(c^d)$ due to the fact that

$$c^d \equiv (m^e)^d \equiv m \,(\text{mod} \, n).$$

However, a number of possible attacks on the RSA algorithm have been described in the past years [2] such as factorization of n.

2.2 The Discrete Logarithm Problem

The next type of public key cryptography protocols is built on the discrete logarithm problem. The discrete logarithm problem was described by Diffie and Hellman in 1976. El Gamal described in 1984 the DL (Discrete Logarithm) public key encryption and signature system. The DL algorithm uses a set of public domain parameters: p, q, g. In this set p is a prime, q is a prime divisor of $p - 1$. g has order q and is selected in $[1, q - 1]$.

The discrete logarithm problem is built on the fact [1, p. 9] that

$$y = g^x \,\text{mod}\, p,$$

where x is a private key. The problem is to determine x from y.

2.3 Elliptic Curve Cryptography

Elliptic Curve Cryptography (ECC) is one of the newest cryptographic systems. Though elliptic curves have been known for the last 150 years it was first discovered to be used in public key cryptography in 1985. Since then ECC has been used for this purpose and is more interesting because it is substantially smaller than RSA [1, p. 19] but definitely more secure. Elliptic Curve systems are built on finite cyclic groups. An Abelian group $(G, *)$ is defined as a set G with a binary operation $* : G x G \rightarrow G$. The Abelian group

G match the following properties [1, pp. 11–12]: *Associativity, existence of an identity, existence of inverses and commutiativity*. In ECC the group operations are normally addition (+) or multiplication (·). The identity element of addition is 0. The additive inverse value of a group b is $-b$. In the multiplicative group the identity element is normally 1. The inverse of the multiplicative group b is b^{-1}. A group is finite if G is a finite set: $\{0, 1 \ldots p - 1\}$ and the number of elements in the group G is the *order of G*. A *finite field* is defined as $(F_p, +, \cdot)$. If G has an element g of n order then g is a generator of G which then is a *cyclic multiplicative group*.

The mathematical problem is the *discrete logarithm problem* and can be shown as extracting a private key from a public key

$$y = g^x,$$

where x is a private key and y is a public key. g is a multiplicative cyclic group and a part of the domain parameters. Also the order n is a domain parameter and the private key x is careful selected in $[1, n - 1]$.

2.4 Non-Adjacent Form (NAF) and τ-adic Non-Adjacent Form (TNAF)

An elliptic curve E over F_p is an equation defined as

$$E : y^2 = x^3 + ax + b.$$

The calculation of a public key using a private key can be done with the NAF (Non-Adjacent Form) [1, p. 98] method for point multiplication. NAF represents the private key k as a signed digit representation

$$k = \sum_{i=0}^{l-1} k_i 2^i \quad \text{where} \quad k \in \{0, \pm 1\}.$$

This representation is guaranteed to be unique, and has an average density of non-zero digits around 1/3 of the length. As an input the NAF method takes a private key k and a point P on the elliptic curve. The point p is a base point which is a public known point defined according to the field size. After calculating the representation the NAF method uses k to calculate a new point on the elliptic curve. The outcome of the NAF representation algorithm is now used for calculating a new point on the elliptic curve. This is done as follows: If k_i is 0 $Q \leftarrow 2Q$. If $k_i = 1$ then first $Q \leftarrow 2Q$ and $Q \leftarrow Q + P$.

If $k_i = -1$ then first $Q \leftarrow 2Q$ and $Q \leftarrow Q - P$. The calculation $Q \leftarrow 2Q$ is called point doubling and is fairly heavy algorithm which consists of 15 steps.

Anomalous binary curves, also known as Koblitz curves, come in two equations:

$$E_0 : y^2 + xy = x^3 + 1,$$

$$E_1 : y^2 + xy = x^3 + x^2 + 1.$$

The big advantage of Koblitz curves is that point doubling can be avoided. A Koblitz curve E_a uses a cofactor h and a prime n and if $nh = \#E_a(F_{2m})$, where h is 4 if $a = 0$ and h is 2 if $a = 1$. Furthermore the Koblitz curve can define the Frobenius map [8] $\tau : E_a(F_{2m}) \rightarrow E_a(F_{2m})$ defined by

$$\tau(\infty) = \infty, \quad \tau(x, y) = (x^2, y^2).$$

This means that the Frobenius map is simple and fast to compute since squaring can be done in hardware near manner.

The Koblitz curve uses a τ-adic non-adjacent form (TNAF) [1, p. 116] instead of NAF explained above. The TNAF algorithm also produces a representation of private key $k : u_i \in \{0, \pm1\}$. The TNAF (k) gives also a unique representation of k. If the length of the TNAF representation has the length l, then the average density of non-zero digits will be $l/3$. The TNAF algorithm can also guarantee that a non-zero digit is followed by zero. The digits produced in the main algorithm of TNAF are calculated by repeatedly dividing k by τ and τ^2 in the following manner:

Let $\alpha = r_0 + r_1\alpha$

If r_0 is even and α is divisible by τ then:

$$\alpha/\tau = (r_1 + \mu r_0/2) - (r_0/2)\tau.$$

Only if $r_0 \equiv 2r_1 \pmod 4$ then α is divisible by τ^2.

The TNAF algorithm is a perfect match for *point multiplication* in Elliptic Curve Cryptography in order to create a public key multiplied with a private key. The TNAF algorithm returns a representation of the private key $k \in \{0, \pm1\}$. First a point Q is set to ∞. Now the sequence above is run through by first setting $Q = \tau Q$. If and only if $u_i = 1$ then $Q \leftarrow Q + P$. If and only if $u_i = -1$ then $Q \leftarrow Q - P$. The outcome is a public key Q.

Participants:

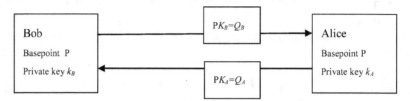

Figure 1 Diffie–Hellman key exchange algorithm (ECC version).

2.5 The Diffie–Hellman Algorithm – The ECC Version

The Diffie–Hellman public key exchange algorithm is a simple protocol [3, pp. 8–10] using exchange of public keys in order to obtain a common shared which can be used in a symmetric cryptographic system.

As stated above the Diffie–Hellman public key exchange is absolutely well fitted to the use of Elliptic Curve Cryptography. The simplicity of the protocol can be seen in Figure 1.

Now Bob wants to make a key exchange with Alice in order to obtain a shared secret. Bob multiplies his private key k_B with the basepoint P. Outcome is a new Point Q_B. This point is now sent to Alice. Receiving this point Alice now does the same as Bob multiplying the private key k_A with the basepoint P resulting in a new point Q_A. Q_A is then sent to Bob. The common shared secret for Bob is $k_B Q_A$ and for Alice $k_B Q_A$, which is the same key. Now this key can be used in a symmetric algorithm as for instance AES (Advanced Encryption System)

As it can be seen, this algorithm is very easy and only requires two communication steps in order to obtain a shared secret.

The problem with this algorithm is that the algorithm is open to Man-in-the-Middle attacks [4] where a third person Eve acts on behalf of Bob exchanging key with Alice and acts as Alice exchanging key with Bob.

2.6 The ECMQV Algorithm

The Man-in-the-Middle problem can be solved with a three-pass protocol as for instance the ECMQV (Elliptic Curve Menezes Qo Vanstone) algorithm. This protocol is using long-term static key pairs [3, p. 10]. Furthermore it is required to calculate two key pairs for each participant in the communication path. As the protocol is a three-pass algorithm it requires three

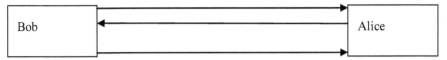

Figure 2 The ECMQV algorithm.

communication paths to obtain a shared secret. The protocol can be seen in Figure 2.

The algorithm has the following calculating sequences:

1. Bob calculates A, R_B (ephemeral key pair) which is sent to Alice
2. Alice sends B, R_A, $t_A = MACk_1(2, B, A, R_A, R_B)$ to Bob
3. Bob sends $t_B = MACk_1(3, A, B, R_B, R_A)$ to Alice.

This protocol is far heavier and requires extra network bandwidth and time to obtain the shared common.

2.7 TLS (Transport Layer Security)

The goal of TLS is to setup private keys for communication in an insecure network. TLS version 1.2 was standardized in August 2008.

The TLS handshake protocol involves a five step communication [5, p. 35] flow between server and client. The protocol can be seen in Figure 3.

The five steps start with a Client Hello sent to the server. This message contains information about the version number, some random generated data to be used to generate a master secret, a session ID, information about the cipher suite and the compression algorithm.

The server responds with a Server Hello. This message includes the chosen version number. A Server Random value is generated which is used in the master secret. The session ID is chosen: Either a new Session ID is created or a session ID is resumed, or a null value. The strongest possible cipher suite will be selected and at last the compression algorithm is selected.

The server sends a server certificate (server's public key) to the client along with a client certificate request. This part of the communication is ended with a Server Hello Done message.

Now the client responds to the server by returning a client certificate. After calculating the premaster secret the client sends a client key exchange to the server. Both the client and server will compute the master secret locally and derive the session keys from it. The Change Cipher Specification message

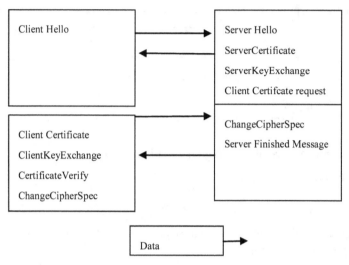

Figure 3 TLS handshake protocol.

tells the server that the following messages will be encrypted. The client ends this part by sending a client finished message.

The server's final response to the client is a Change Cipher Specification message and a server finished message. As it can be seen, this is a very heavy weight protocol.

2.8 AKE (Authenticated Key Exchange) Protocols

Huang and Cao have proposed an ID based AKE protocol [6]. The protocol is based on bilinear pairing using two cyclic groups of prime order $q (e = GxG \rightarrow GT)$. $P \in G$ is the generator of group G.

Furthermore the protocol uses three hash functions. There are two participants, A and B. When A is initiating the communication, A chooses an ephemeral private key $x \in Z$. Next A computes an ephemeral public key $X = xP$. X is sent to B. B is doing similarly and sends $Y = yP$ to A. When receiving B verifies that $X \in G$. B computes:

$$Z1 = e(X + QA1; yZ + dB1),$$

$$Z2 = e(X + QA2; yZ + dB2),$$

$$Z3 = yX \quad \text{and}$$

$$SK = H(Z1; Z2; Z3; \text{sid}), \quad \text{sid} = (X; Y; A; B).$$

B keeps *SK* as a session key. A performs similar calculation to obtain the session key.

Jooyoung Lee and Je Hong Park have proposed a new AKE protocol [7] called the NAXOS+ protocol which is a modified NAXOS protocol as proposed by LaMacchia, Lauter and Mityagin.

The NAXOS+ protocol involves a CA (Certificate Authority). The CA checks if the public static key is contained in G^*. Via a certificate a participant obtains knowledge about each other.

For each session an ephemeral public/private key pair is generated. In order to obtain a MAC of the message a three-pass algorithm NAXOS+C.

2.9 AES (Advanced Encryption Standard)

The Advanced Encryption Standard (AES) became standard in 2001 and is described in the FIPS-197 document from NIST (National Institute of Standards and Technology).

At the time AES is considered to be one of the most secure symmetric algorithms.

AES is non-feistel algorithm; not the same algorithms are used for encryption and decryption.

As in normal block cipher, AES can be used in several modes [8, p. 151]: CBC, EBC, CFB, CTR and OFB.

The AES algorithm is using a fixed block size of 128 bits and uses different key sizes of 128, 192 or 256 bits depending on the security level. In AES four operations are used: AddRoundKey, SubBytes, ShiftRows and MixColumns. The order of four operations follows a well-known described scheme in the main algorithm consisting of rounds: In the initial round AddRoundKey is performed. In the following rounds SubBytes, ShiftRows, MixColumns and AddRoundKey are performed. In the last round only Sub-Bytes, ShiftRows and AddRoundKey are performed. If the key size is 128 bits 10 rounds are executed, if the key size is 192 bits the number of rounds is 12 and finally if the key size is 256 bits the number of rounds is 14.

2.10 HMAC (Hash-based Message Authentication Code)

The HMAC is from 1997 and is publish in the RFC 2104 document. According to the document the HMAC algorithm can be used with any iterative cryptographic hash functions. At the time being there exists two standard implementations: SHA-1 and SHA-2. SHA stands for Secure Hash Standard.

SHA-3 implementation is to be decided in 2012. HMAC consists of a cryptographic hash function H and a secret key K. Also two fixed strings are used; ipad (inner) and opad (outer). Ipad contains the byte 0x36 B times, where B is the length of K. *opad* contains the byte 0x5C B times.

The HMAC is now computed [9, p. 3] over some text *aText*:

$$H(K\ XOR\ opad), H(K\ XOR\ ipad, aText)$$

The SHA-2 cryptographic function is rather a family of functions: SHA-224, SHA-256, SHA-384, and SHA-512, where the number denotes the digest length in bits. The SHA-2 algorithms were published in 2001 in FIPS PUB 180-2 designed by the NIST (National Institute of Standards and Technology). The SHA-2 is considered substantially more secure than the SHA-1.

3 Secure Plain Diffie–Hellman Algorithm

In this section we introduce the Secure Plain Diffie–Hellman algorithm (SPDH). The SPDH algorithm can be used where an AuC (Authentication Center) is present.

3.1 Introduction to the Secure Plain Diffie–Hellman Algorithm

The algorithm addresses the following issues:

- It is a lightweight two pass algorithm.
- It uses basic Diffie–Hellman (Elliptic Curve) approach.
- It is secured against Man-in-the-Middle attacks.
- It is secured against replay attacks.
- Security relies on ECC, HMAC and AES.
- It can use personal IDs – not only hardware IDs.
- It can be routed on a network without being exposed.
- Initial keys ready from start.
- Initial keys can be renewed.
- Secure transport of Diffie–Hellman keys.
- It uses well-known and tested algorithms.

3.2 Secure Plain Diffie–Hellman Algorithm

INPUT: A basepoint P, private key k, a private key x_{AES}, a private key y_{HMAC}, a personal ID *pid*, Timestamp ts

PARTICIPANTS: A, B and Authentication Center (AuC)

OUTPUT: A shared secret G

1. A calculates $P_{(A)k}$
2. A calculates $\text{HMAC}(\text{AES}(PA_{(A)k} + pid_A + ts))$ using $x_{AES(A)}$ and $y_{HMAC(A)}$
3. $\rightarrow$ AuC Checks (2) (Recalculates HMAC and checks that ts is newer than listed and reliable)
4. AuC unpack $\text{HMAC}(\text{AES}(PA_{(A)k} + pid_A + ts))$ using $x_{AES(A)}$ and $y_{HMAC(A)}$
5. AuC recalculates $\text{HMAC}(\text{AES}(PA_{(A)k} + pid_A + ts))$ using $x_{AES(B)}$ and $y_{HMAC(B)}$
6. $\rightarrow$ B Checks (5) (Recalculates HMAC and checks that ts is newer than listed and reliable)
7. B unpack (6) using $x_{AES(B)}$ and $y_{HMAC(B)}$
8. B calculates $P_{(B)k}$ and stores $P_{(A)k}$
9. B calculates $\text{HMAC}(\text{AES}(P_{(B)k} + pid_B + ts))$ using $x_{AES(B)}$ and $y_{HMAC(B)}$
10. $\rightarrow$ AuC checks (9) (Recalculates HMAC and checks that ts is newer than listed and reliable)
11. AuC unpack $\text{HMAC}(\text{AES}(P_{(B)k} + pid_B + ts))$ using $x_{AES(B)}$ and $y_{HMAC(B)}$
12. AuC recalculates $\text{HMAC}(\text{AES}(P_{(B)k} + pid_B + ts))$ using $x_{AES(A)}$ and $y_{HMAC(A)}$
13. $\rightarrow$ A Checks (12) (Recalculates HMAC and checks that ts is newer than listed and reliable)
14. A unpack (13) using $x_{AES(A)}$ and $y_{HMAC(A)}$ and stores $P_{(B)k}$
15. Common shared secret is $P_{(A)k} P_{(B)k}$

It is assumed that the private k is generated by the participant, the private keys for A: $x_{AES(A)}$, $y_{HMAC(A)}$ and $x_{AES(B)}$, $y_{HMAC(B)}$ are delivered from authentication center (AuC) in the initial phase during setup. The personal ID is also generated by the participant and is well known by other participants.

As it can be seen from this algorithm not even the AuC can calculate the common shared secret $P_{(A)k} P_{(B)k}$ since only the participants themselves know the private key k.

The key exchange is secured further. The HMAC algorithm solves the problem of malicious altering of the key exchange and will also prevent replay attacks. The message uniqueness is guaranteed by using Timestamp ts.

The algorithm has some further requirements. The following is assumed:

1. A and B are participants in the system and the authentication procedure between AuC and the participants are done correctly.
2. The keys $x_{AES(A)}$, $y_{HMAC(A)}$ and $x_{AES(B)}$, $y_{HMAC(B)}$ are created in a challenge-response manner with AuC.
3. En elliptic curve base point is created along with other domain parameters.
4. A *pid* is selected.
5. A private key k for each participant is selected.
6. The Timestamp ts is created locally for both participants A and B plus the AuC.
7. A, B and AuC maintains a list of timestamps used for checking that a message is new and never received before. The Timestamp is also encrypted.
8. The ECC, HMAC and AES software are installed.

3.3 Network Topology – Involving More Than One AuC

Since the content is secret during transport on the network involving more than one AuC in a network (roaming) does not impose further problems. It is assumed that the involved AuC know and trust each other and have exchanged encryption/decryption keys. In that way a single AuC can transport encrypted information without having knowledge about the shared secret. It is not demanded that the transport of the content is secured previously for instance using SSL/TLS.

The main flow can be shown as described below.

3.4 The Routed Secure Plain Diffie–Hellman Algorithm

INPUT: A basepoint P, Private key k, a private key x_{AES}, a private key y_{HMAC}, a personal ID *pid*, Timestamp ts, a network routing table rt, receiver info ri.

PARTICIPANTS: A, B and Authentication Centers ($AuC_i^{i=0}$, AuC_{i+1} ... AuC_{i+n})

OUTPUT: A shared secret G

Note: The input parameter rt is a routing table. This routing table routes the Diffie–Hellman setup information through a network that can change depending on transmission time, cost or other parameters. The routing table is typically created as a matrix in two or more dimensions. The ri input parameter is a unique information about the receiver of the information (i.e. roaming information).

1. A calculates $P_{(A)k}$
2. A calculates $HMAC(AES(PA_{(A)k} + pid_A + ts))$ using $x_{AES(A)}$ and $y_{HMAC(A)}$
3. A sends(2) to AuC_0
4. AuC_i Checks (3) (Recalculates HMAC and checks that ts is newer than listed and reliable)
5. AuCi unpack $HMAC(AES(PA_{(A)k} + pid_A + ts))$ using $x_{AES(A)}$ and $y_{HMAC(A)}$
6. AuC_i recalculates $HMAC(AES(PA_{(A)k} + pid_A + ts + rt + ri))$ using $x_{AES(AuCi)}$ and $y_{HMAC(AuCi)}$
7. AuC_i Sends (6) to AuC_{i+1}
8. AuC_{i+1} (Recalculates HMAC and checks that ts is newer than listed and reliable)
9. AuC_{i+1} unpack (8) using $x_{AES(AuCi)}$ and $y_{HMAC(AuCi)}$
10. AuC_{i+1} recalculates $HMAC(AES(PA_{(A)k} + pid_A + ts + rt + ri))$ using $x_{AES(AuCi+2)}$ and $y_{HMAC(AuCi+2)}$
11. AuC_{i+1} sends (10) toAuC_{i+2}
12. For length of rt: Step 4 to 12
13. AuC_{i+n} Checks (12) (Recalculates HMAC and checks that ts is newer than listed and reliable)
14. AuC_{i+n} unpack $HMAC(AES(PA_{(A)\,k} + pid_A + ts + rt + ri))$ using $x_{AES(AuCi+n)}$ and $y_{HMAC(AuCi+n)}$
15. AuCi+n recalculates $HMAC(AES(PA_{(A)k} + pid_A + ts))$ using $x_{AES(B)}$ and $y_{HMAC(B)}$
16. AuC_{i+n} reads ri (now empty) and sends (15) to B
17. B checks (16) (Recalculates HMAC and checks that ts is newer than listed and reliable)
18. B checks that pid_A is well known and accepted
19. B calculates $P_{(B)k}$ and stores $P_{(A)k}$
20. B calculates common shared secret $G = (P_{(A)k}\ P_{(B)k})$
21. B calculates $HMAC(AES(P_{(B)k} + pid_B + ts))$ using $x_{AES(B)}$ and $y_{HMAC(B)}$
22. B sends (21) to AuCi *in reverse order* AuC_{i+n} to AuC_0
23. AuCi checks (22) (Recalculates HMAC and checks that ts is newer than listed and reliable)

24. AuCi unpack HMAC(AES($P_{(B)k} + pid_B + ts$)) using $x_{AES(AuCi+1)}$ and $y_{HMAC(AuCi+1)}$
25. AuCi recalculates HMAC(AES($P_{(B)k} + pid_B + ts + rt + ri$)) using $x_{AES(A)}$ and $y_{HMAC(A)}$
26. AuCi sends (25) to A
27. A checks (26) (Recalculates HMAC and checks that ts is newer than listed and reliable)
28. A checks that pid_B is well known and accepted
29. A unpack (27) using $x_{AES(A)}$ and $y_{HMAC(A)}$ and stores $P_{(B)k}$
30. Common shared secret is $G = (P_{(A)k} \; P_{(B)k})$

4 Experimental Results and Analysis

In order to test the performance of the algorithm, the system consisting of Elliptic Curve Cryptography, HMAC and AES has been implemented in C++.

4.1 Software Implementation

The software implementation of the Elliptic Curve Cryptography is built in three layers: (1) a basic layer for operations on bit strings, (2) a second layer for handling fields; and (3) the third layer is handling Elliptic Curve Cryptography operations. The Elliptic Curve Cryptography implementation is an anomalous binary curve also known as a Koblitz curve.

HMAC is implemented as described in FIPS 180-2 (Federal Information Processing Standards Publications) from National Institute of Standards and Technology (NIST) and is the SHA-256 (Secure Hash Algorithm) version.

The flow is as described in Section 3.2. First A is calculating a point on the curve using the private key. Next a message encryption of the point along with the personal ID and the timestamp is done. After this the HMAC value is calculated. AuC recalculates the HMAC value and decrypt the message. After this the AuC encrypt the message with B's key and pass the message on to B along with a newly created HMAC value.

B recalculates the HMAC value and unpacks the message.

Now B creates a response message with a calculated point using B's private key along with the personal ID of B and a timestamp. The AuC repeats the decryption-encryption of the message from B and pass the message on to A.

The common share between A and B is $QK_A \; K_B$.

Table 1 Operation times.

Operation	Time [mS]
AES Encryption	0.014
AES Decryption	0.015
HMAC Hashcode	0.003
ECC Point calculation	105.0

4.2 Performance Analysis

It is well known that symmetric algorithms are faster than asymmetric algorithms. In another word the point calculation in ECC is slower than the AES encryption/decryption and calculation of the HMAC value.

As it can be seen above the asymmetric algorithm is only calculated on end-points (participants A and B). Furthermore, the participants A and B have to calculate AES and HMAC. The AuC calculates only the symmetric algorithms and therefore the data transport between the A and AuC and between B and AuC is fast.

The Routed SPDH algorithm will in the same way benefit from a fast algorithm; still only the end-point participants have to calculate the asymmetric algorithm so load on the shared AuCs is limited.

The speed in the system has been tested using an Intel® Core™ i3 CPU @ 2.27 GHz. In this performance test, the network time is not included.

The test results are shown in Table 1. These values will in real systems first of all depend on the calculation power of the participants A and B. Second, the load on the AuC will influence on calculation time. Third, the network load will also influence on calculation time.

5 Verification

5.1 Verification of the SPDH Algorithms

Both the SPDH algorithm and the Routed SPDH algorithm which is an extended version of SPDH algorithm have been verified.

The SPDH algorithms have been tested using ProVerif [10]. ProVerif is an automatic cryptographic protocol verifier, which is able to handle cryptographic primitives and also for instance public key cryptography. Especially Diffie–Hellman key agreements and hash functions can be documented using this tool.

The algorithms have been tested for passive attacks and active attacks. In the passive attack the attacker listen to the communication (eavesdropping).

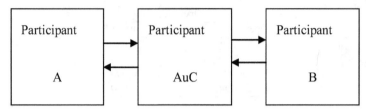

Figure 4 Participants in SPDH.

Figure 5 Participants in Routed SPDH.

In active attacks, as for instance Man-in-the-Middle attacks the attacker tries to alter data.

For the SPDH algorithm the setup in ProVerif is as shown in Figure 4.

For the Routed SPDH algorithm the setup in ProVerif is as shown in Figure 5.

It is well known that there is a vulnerability in the Diffie–Hellman algorithm, where the attacker can perform a Man-in-the-Middle attack simply by intercepting the communication initiated by A and B. In this attack type the attacker, E, intercepts the public key from A. Then E transmits E's own public key to B. B then transmits his public key to E without knowing that is E and not A, E also intercept the key to B.

No mechanism in the Diffie–Hellman protects against replay attacks. A replay attack is to done by retransmit valid data on the network that have been captured previously [11]. By using a timestamp retransmitted data can be discarded.

5.2 Verification Results

The results from the ProVerif test are presented in Table 2.

5.3 Limitations

AuCs must be reliable and trusted.

Table 2 ProVerif test results.

Algorithm	Eavesdropping (Passive)	Replay attack (Active)	Man-In-The-Middle attack (Active)
Diffie–Hellman algorithm	Secure	Vulnerable	Vulnerable
SPDH algorithm	Secure	Secure	Secure
Routed SPDH algorithm	Secure	Secure	Secure

6 Conclusion

A new and more secure variant of the original Diffie–Hellman algorithm has been introduced.

The Secure Plain Diffie–Hellman algorithm has the advantage of the original Diffie–Hellman algorithm, which is a fast two-pass communication algorithm. The secure Plain Diffie-Hellman algorithm is based on Elliptic Curve Cryptography and uses standard implementations of AES and HMAC to enhance security.

The Secure Plain Diffie–Hellman algorithm is protected against Man-in-the-Middle attacks, eavesdropping and replay attacks.

The proposed Secure Plain Diffie–Hellman algorithm can be used in numerous systems where a trusted central unit is present. This could for instance be in the mobile telephone system, in wireless systems, in secure sensor systems or military field units.

The Secure Plain Diffie–Hellman algorithm is especially well suited for mobile communication systems where the need is to achieve an endpoint-to-endpoint secure communication. But also in situations where ad-hoc network can be created the Secure Plain Diffie–Hellman algorithm can be used to provide endpoint-to-endpoint secure communication.

The Secure Plain Diffie–Hellman algorithm has also been presented in a routed version. The routed version is well suited in mobile communication systems where endpoint-to-endpoint users are using different providers or are roaming.

The performance test has shown that the use of an Authentication Center (AuC) will not influence substantially on the general performance since the operations on the AuC are only symmetric operations and thereby fast.

References

[1] Hankerson et al. Guide to Elliptic Curve Cryptography. Springer, 2004.

[2] D. Boneh. Twenty years of attacks on the RSA cryptosystem. Notices of the American Mathematical Society (AMS), 46(2):203–213, 1999.

[3] F. Blake (Ed.). Advances in Elliptic Curve Cryptography. Cambridge University Press, 2005.

[4] Mario Cagalj, Srdjan Capkun, and Jean-Pierre Hubaux. Key agreement in peer-to-peer wireless networks. IEEE (Special Issue on Cryptography and Security). bibitem5. RFC 5246, 2008.

[5] Hai Huang and Zhenfu Cao. An ID-based authenticated key exchange protocol based on bilinear Diffie–Hellman problem. Department of Computer Science and Engineering, Shanghai Jiaotong University, ASIACCS, 2009.

[6] Jooyoung Lee and Je Hong Park. Authenticated key exchange secure under the computational Diffie–Hellman assumption. The Attached Institute of Electronics and Telecommunications Research Institute, Korea, IACR, 2008.

[7] W. Trappe and L.C. Washington. Introduction to Cryptography with Coding Theory (second edition). Pearson, 2006.

[8] RFC 2104.

[9] http://www.proverif.ens.fr/.

[10] Priyanka Goyal, Sahil Batra, and Ajit Singh. A literature review of security attack in mobile ad-hoc networks. International Journal of Computer Applications, 9(12):11–15, November 2010.

Biographies

Henrik Tange received the B.Eng (export engineer) from the Copenhagen University College of Engineering in 1999 and the M.Sc. in Communication Network specializing in Security from Aalborg University in 2009. Since 2009 he has been a PhD student at Aalborg University. Since 2000 he has been teaching at Copenhagen University College of Engineering.

Birger Andersen is a professor at Copenhagen University College of Engineering, Denmark, and director of Center for Wireless Systems and Applications (CWSA) related. He received his M.Sc. in Computer Science in 1988 from University of Copenhagen, Denmark, and his Ph.D. in Computer Science in 1992 from University of Copenhagen. He was an assistant professor at University of Copenhagen, a visiting professor at Universität Kaiserslautern, Germany, and an associate professor at Aalborg University. Later he joined the IT Department of Copenhagen Business School, Denmark, and finally Copenhagen University College of Engineering. He is currently involved in research in wireless systems with a focus on security.

Impact of Constraints on the Complexity and Performance of Channel Assignment in Multi-Hop Wireless Networks

Chetan Nanjunda Mathur, M.A. Haleem, K.P. Subbalakshmi and
R. Chandramouli

*Department of Electrical and Computer Engineering, Stevens Institute of
Technology, Hoboken, NJ 07030, USA; e-mail: mouli@stevens.edu*

Abstract

In this paper we systematically study several channel assignment problems
in multi-hop ad-hoc wireless networks in the presence of several constraints.
Both regular grids and random topology models are considered in the analysis. We identify three fairness constraints (unfair, fair, and 1-fair), Signal
to Interference Ratio (SINR) constraint (to measure the link quality) and
balance constraint (for uniform assignment) and study their impact on the
complexity of the channel assignment problems. Note that these constraints
have an impact on the network capacity, lifetime and connectivity.

Although optimal channel assignment for links in a multi-hop wireless
network has been shown to be NP complete, the impact of fairness, link
quality and balance constraints on the hardness of channel assignment problems is not well studied. In this paper, we show that a class of unfair SINR
constrained channel assignment problems can be solved in polynomial time.
We show that when fairness is desired the channel assignment problems are
NP Complete. We propose two heuristic algorithms that provide 1-fair and
fair channel assignments, comment on their complexity and compare their
performance with optimal solutions.

Keywords: scheduling, set covering, graph coloring, wireless ad-hoc networks, NP.

Journal of Cyber Security and Mobility, Vol. 1, 161–187.

1 Introduction

Based on the number of wireless hops, wireless networks can be classified into last-hop and multi-hop networks. Last hop networks refer to wired networks whose last hop is wireless. These types of networks are also called infrastructure networks, the cellular network being an example. On the other hand if all the hops of a wireless network are comprised of wireless links, the network is called a multi-hop wireless network. An example for such networks is the ad-hoc network. In this paper we concentrate on multi-hop wireless networks.

As all the nodes in a multi-hop network share a common wireless communication medium, some type of coordination is desired to avoid collisions or simultaneous channel access by two or more devices. Most of the literature on medium sharing techniques can be classified into two types: random access techniques and channel assignment techniques. Random access techniques, as the name suggests, use some type of randomized protocols to access the channel in such a way that collisions are minimized. Examples of random access techniques include the well known Aloha protocol [11] and the Carrier Sense Multiple Access with Collision Avoidance (CSMA/CA). Channel assignment techniques on the other hand, use orthogonal sub-channels as the key to achieving simultaneous communication and mitigate mutual interference. This is used both in networks with global co-ordination and distributed co-ordination. The channel assignment techniques in wide use are Time Division Multiple Access (TDMA), Frequency Division Multiple Access (FDMA), Code Division Multiple Access (CDMA), Orthogonal Frequency Division Multiplexing (OFDM) and some combinations of these [11]. Therefore by the term channel we mean, a time frame in TDMA, a segment of allocated spectrum in FDMA, or a collection of orthogonal codes in CDMA. The term sub-channel thus implies time slots of a time frame, frequency channels within the segment of the spectrum or a code in the collection of orthogonal codes. For the analysis in the paper we assume time slot assignment.

Depending on what is being scheduled, channel assignment can be of two types. When nodes are assigned to different sub-channels [4, 7, 10, 14], it is referred to as node scheduling/assignment. Node scheduling is usually referred to as broadcast scheduling. Similarly, link scheduling is when each of the links in the network are assigned sub-channels. Link scheduling is also referred to as point to point scheduling [1, 2, 6, 12]. Although node and link scheduling problems differ in terms of number of elements to be scheduled

Table 1 Notations used.

V	Set of all nodes (vertices of G)
$e_{i,j}$	Link formed when node i transmits to node j
	Note that the existence of $e_{i,j}$ does not imply $e_{j,i}$
E	Set of all possible links in the network (edges in G)
E_t	Set of links that are active at time t, where $t > 0$
$SIR(e_{i,j}, t)$	Signal to interference ratio of edge $e_{i,j}$ at time t
T	Total number of active sub-channels (time-slots)
S_i	Set of sub-channels assigned to link e_i

and computational complexity, in general a solution to link scheduling can be transformed to node scheduling by transforming the nodes in the given graph into links and by applying link scheduling on it. Some of the work that propose algorithms to achieve both broadcast scheduling and point to point scheduling are [13, 16]. In our current work we only consider link scheduling problems. Hence in the rest of the paper by channel assignment we refer to link scheduling. We identify three constraints in channel assignment: (a) signal to interference noise ratio constraint, (b) fairness constraint, and (c) balance constraint. A formal definition of these constraints are given in later sections. We note that these constraints impact the network capacity, connectivity, lifetime and traffic load. In addition to these we also discuss atomic constraints that are dictated by the physical layer of the nodes.

2 Network Model

To formulate the channel assignment problem in a multi-hop wireless network, we map a given wireless network to a graph $G = (V, E)$, where the vertices, V, represent the wireless devices (nodes) and the edges, E, represent the transmission links between these devices. We capture the multi hop wireless networks that occur in practice using graphs as described later in this section. The network graphs used in the experimental section are generated using this model. Some of the notations used in the rest of the paper are summarized in Table 1.

2.1 Random Network Model

In this case, the nodes are randomly placed in a given network area. A fixed power level is allocated to each of the nodes. We define the parameter $R_{\max}$ of the network as the length of the longest active edge. Similarly $R_{\min}$ is the

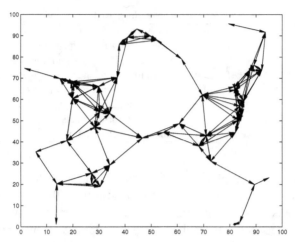

Figure 1 Example of random network with area $A = 100 \times 100$ square meters, 49 nodes and range $R_{\max} = 20$ meters, $R_{\min} = 1.362$ meters.

length of the smallest active edge in the network. An example of random network model is shown in Figure 1.

2.2 Grid Based Network Model

In the grid based network model we assume that the nodes are placed on a rectangular grid. In this model all the edges formed are of same length and all transmitting nodes have same range. Hence the maximum transmission range $R_{\max}$ is same as the minimum transmission range $R_{\min}$. An example of grid based network model for a 49 node network is shown in Figure 2. This type of model is also used in, for example, [5].

3 Channel Assignment Constraints

3.1 Atomic Constraint

An atomic constraint is a symmetric relationship between two vertices or two edges in a graph [13]. Two vertices/edges that are mutually constrained cannot be scheduled in the same sub-channel. In [13], the author identifies atomic constraints and models scheduling/assignment problems as a constraint set, A_c, over the atomic constraints. For the sake of clarity, in Table 2 we summarize the 3 atomic link constraints that are relevant to our work.

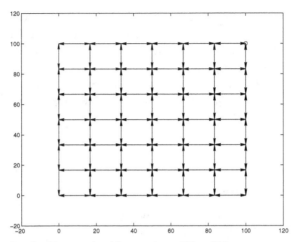

Figure 2 Example of grid network with area $A = 100 \times 100$ square meters, 49 nodes and range $R_{\min} = R_{\max} = 20$ meters.

Table 2 Atomic Link Constraints

Constraint	Implication
$(E, 0, tt)$	$\nexists e_{i,j}, e_{i,k}$ s.t. $\{e_{i,j}, e_{i,k}\} \subseteq E_t, t \in \{1, \dots, T\}$
$(E, 0, rr)$	$\nexists e_{j,i}, e_{k,i}$ s.t. $\{e_{j,i}, e_{k,i}\} \subseteq E_t, t \in \{1, \dots, T\}$
$(E, 0, tr)$	$\nexists e_{i,j}, e_{k,i}$ s.t. $\{e_{i,j}, e_{k,i}\} \subseteq E_t, t \in \{1, \dots, T\}$

For example the TDMA link scheduling/assignment problem can be modeled as a constraint set $A_c(TDMA) = \{(E, 0, tt), (E, 0, rr)\}$. $(E, 0, tt)$ means that in any time slot a node cannot perform multiple simultaneous direct transmissions to more than one receiving node. $(E, 0, rr)$ means that a receiving node cannot directly receive simultaneously from two different transmitting nodes in a given time slot.

3.2 SINR Constraint

Although link schedules that satisfy atomic constraints avoid collisions in the wireless network, they do not impose any kind of guarantee on the quality of received signal at each receiver. To assure a minimum quality at each receiver in the network, we define a constraint on the signal to interference plus noise ratio (SINR). The SINR constraint simply says that the assignment should guarantee a desired minimum SINR requirement (say γ_{th}) for all the links assigned in all of its sub-channels. While the optimal assignment of links and power control to satisfy this constraint requires the consideration of the "physical model" of the system, a simplified analysis can be carried out with a conservative approach leading to the so called "protocol model" [5]. We now define the SINR constraint in the physical model which is then simplified to distance constraint in the protocol model.

Let E_t be the set of all links assigned to the sub-channel t, then under the physical model the scheduling/assignment is said to satisfy the minimum SINR constraint if

$$SINR(e_{ij}, t) \geq \gamma_{th}, \forall t \in \{1, \ldots, T\}, \forall e_{ij} \in E_t \tag{1}$$

3.3 Distance Constraint

While the physical model simply states the constraint, the protocol model discussed below provides a simplified method to impose the constraint.

If link e_i is assigned sub-channel t (Figure 3) i.e., $e_i \in E_t$, then it is said to satisfy the minimum distance ratio constraint [5], if $\forall e_j \in E_t - e_i$,

$$d(Tx(e_j), Rx(e_i)) \geq (1 + \delta_{th})R_{\max}, \tag{2}$$

where $R_{\max}$ is the maximum transmission radius in the network, $Tx(e_j)$ denotes the transmitter of link e_j and $Rx(e_i)$ is the receiver of link e_i. That is, the distance between the receiver of the given link and the transmitters of other active links sharing the same sub-channel t should be larger by a factor of $1 + \delta_{th}$ compared to the maximum transmission radius of the network. Note that for the grid based network model $R_{\max} = R_{\min} = R$.

3.4 Fairness

Depending on how many sub-channels each link is assigned to, channel assignment can be classified as unfair, 1-fair and fair. A channel assignment is called unfair if there is at least one link that is not assigned to any of the sub-channels. This leads to the following definition.

Definition 1 *Let S_i be the set of sub-channels assigned to link e_i, then S_i is represented as*

$$S_i = \{t, e_i \in E_t \forall t \in \{1, \ldots, T\}\} \tag{3}$$

An assignment is said to be unfair if

$$\exists i \in \{1, \ldots, |E|\} \text{ s.t. } |S_i| = 0 \tag{4}$$

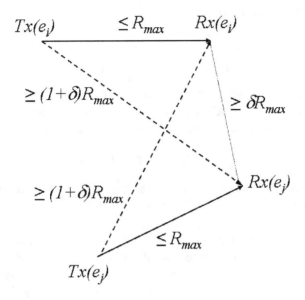

Figure 3 Pairwise distance constraints between two given links e_i and e_j in a wireless network, with distance ratio constraint of δ.

Under such a channel assignment the node degree will reduce and the connectivity of the graph may be lost, thus making routing between some source destination pairs impossible. Nevertheless, a network that prefers "maximum overall throughput" regardless of fairness may adopt an unfair policy.

In order to preserve the connectivity of the underlying communication graph, we need every link to be assigned to at least one sub-channel. This gives rise to the following definition of 1-fairness.

Definition 2 *A channel assignment is said to satisfy the 1-fairness constraint if*

$$|S_i| \geq 1, \ \forall i \in \{1, \ldots, |E|\} \tag{5}$$

Although 1-fair assignment preserves the connectivity of the graph it can be biased. Some of the links could be assigned more sub-channels. Hence, when every link is assigned to exactly same number of sub-channels, such an assignment achieves fairness.

Definition 3 *A channel assignment is said to be fair if,*

$$|S_i| = d, \ \forall i \in \{1, .., |E|\} \tag{6}$$

where d ≥ 1 is a constant.

3.5 Balance

Unfair, 1-fair and fair assignments need not be balanced over sub-channels; some of the sub-channels may be assigned more number of links than the others. To achieve an ideally balanced assignment, it is necessary that every sub-channel contains exactly the same number of active links. That is,

$$|E_{t_1}| = |E_{t_2}|, \forall t_1, t_2 \in \{1, \ldots, T\} \tag{7}$$

otherwise, the channel assignment is said to be unbalanced. Balanced assignment will help in maintaining an acceptable level of load per time slot.

It is easy to see that ideally balanced assignment is not always feasible; one example is when the number of links in the network is prime. Hence, we define a channel assignment to be ϵ_{th} unbalanced if the maximum absolute difference in the number of active links between two sub-channels is not greater than ϵ_{th}. That is,

$$\max(abs(|E_{t_1}| - |E_{t_2}|)) \le \epsilon_{th}, t_1, t_2 \in \{1, \ldots, T\} \tag{8}$$

Note that ϵ_{th} is upper bounded by $|E|$.

4 Link Scheduling/Sub-Channel Assignment Problems

Definition 4 *A Channel Assignment Problem (CAP) is represented as a two-tuple* $(\mathfrak{G}, \mathfrak{C})$, *where*

1. $\mathfrak{G}$ *is a two-tuple* $(\mathbb{G}, \mathbb{P})$ *representing the wireless network with,*
   - $\mathbb{G}$ *being a set of graph representations G of the wireless network.*
   - $\mathbb{P}$ *being a set of network parameters such as geographical area A, set of link gains $\{h\}$ as well as the presence/absence of "power control".*
2. $\mathfrak{C}$ *is a four-tuple* $(\mathbb{A}, \mathbb{F}, \mathbb{D}, \mathbb{E})$ *that represents the set of all constraints*
   - $\mathbb{A}$ *is a finite set of possible atomic constraint sets $\{A_c\}$*
   - $\mathbb{F}$ *is a finite set $\{unfair, 1\text{-}fair, fair\}$ of possible fairness constraints*
   - $\mathbb{D}$ *is a set of quality constraints such as distance constraint $\delta_{th} \ge 0$ or minimum SINR constraint*
   - $\mathbb{E}$ *is a set of all balance constraints $\epsilon_{th} \ge 0$.*

Table 3 Channel assignment problems.

$\mathbb{F}$	$\mathbb{E}$	Channel Assignment Problem
unfair	–	unfair distance constrained assignment
1-fair	$|E|$	1-fair distance constrained assignment
fair	$|E|$	fair distance constrained assignment
fair	ϵ_{th}	fair, balanced and distance constrained assignment

The channel assignment algorithms (CAA) have two basic objectives:

- *maximize the number of links in every sub-channel (maximized total link capacity),*
- *minimize the total number of sub-channels used subject to all the constraints (minimizes total delay).*

The sub-channels we consider are the time slots in a time frame. Depending on fairness, distance and balance constraints we may classify various channel assignment problems as shown in Table 3.

We will now formulate the two different objectives (as given in Definition 4) of a CAA. In doing so, we identify the correspondence of these problems to other well known problems to get an idea on the complexity. Henceforth, we call the problem of maximizing the number of links assigned to a given sub-channel as the *sub-channel assignment problem* and the problem of minimizing the number of sub-channels used in channel assignment as the *sub-channel minimization problem.*

4.1 Sub-Channel Assignment Problem (SAP)

Given a multi-hop wireless network $G=(V, E)$, if a subset of edges $\{e\} \subseteq E$ satisfies A_c and δ_{th}, this set is called a transmission set [1]. A naive way to enumerate all the transmission sets for G is to take all possible combinations of edges in E and test to see if they satisfy the specified constraints. The complexity of this algorithm is

$$\sum_{i=1}^{|E|} \binom{|E|}{i} \qquad (9)$$

which is exponential in $|E|$. The objective of SAP is to find out the largest possible transmission set for a given sub-channel.

Problem 1 The objective of SAP is to assign maximum number of links to a given sub-channel subject to the specified atomic and distance constraints A_c and δ_{th}, respectively.

To analyze the complexity of SAP, we compare it with the well known Max-Clique problem [3]. Given is an instance of SAP, (G, c), where $c \in C$ is a set of given constraints. Algorithm *SAP-to-MaxClique* transforms the objective of SAP to that of finding maximum clique in a graph. Note that in the graph

---

Algorithm 1 *SAP-to-MaxClique*

Input: (G, c) of SAP
Output: $H = (V_a, E_a)$ an instance of $MaxClique$

1: $H(V_a) \leftarrow G(E)$
2: **for** $v_i = 1 : V_a$ **do**
3:      **for** $v_j = 1 : V_a$ **do**
4:          **if** Edges v_i and v_j in G are not mutually constrained by any constraint in c **then**
5:              $H(E_a) \leftarrow H(E_a) \cup (v_i, v_j)$
6:          **end if**
7:      **end for**
8: **end for**

$H = (V_a, E_a)$, vertices of H represent edges of G and any two edges in G that are mutually constrained in $c \in C$ do not have an edge in H between the corresponding vertices.

The complexity of algorithm *SAP-to-MaxClique* is $\Theta(|E|^2)$, which is polynomial in E. As none of the vertices in any complete subgraph of H are mutually constrained by any of the constraints specified by c, they all can be scheduled in the same sub-channel. This makes every complete subgraph of H a transmission set of G. Hence the sub-channel allocation problem is to determine the maximum clique of H. The decision version of this problem is to determine whether a clique of a given size J exists in the graph.

Problem 2 MaxClique decision problem.
Instance: A graph $H=(V, E)$ and a positive integer $J \leq |V|$.
Question: Does H contain a clique of size J or larger, that is, is there a subset $V' \subseteq V$ such that $|V'| \geq J$ and every two vertices in V' are joined by an edge in E?

If J is not a constant significantly smaller than $|V|$, this problem and the corresponding optimization problem are known to be NP complete [15]. Note that a naive way to solve SAP is to enumerate all possible sub-graphs of H and determine if the size of largest complete subgraph is J. The running time of the naive algorithm (*Naive-MC*) to solve the MaxClique is $\Omega(J^2\binom{|V|}{J})$, which is polynomial if J is a constant [3].

4.2 Sub-Channel Minimization Problem (SMP)

Given a multi-hop wireless network $G = (V, E)$, the number of transmission sets is upper bounded by (9). However, for given constraints $c \in \mathcal{C}$ the number of feasible transmission sets may be reduced. The objective of the sub-channel minimization problem is to find a collection of minimum number of transmission sets such that every edge in the network appears in at least one transmission set.

Problem 3 The sub-channel minimization problem (SMP) is to find a *1-fair* or *fair* assignment that minimizes the number of sub-channels, T, subject to the specified atomic, distance and balance constraints A_c, δ_{th}, ϵ_{th} respectively. In other words, minimize T, such that $\bigcup_{t=1}^{T} E_t = E$, A_c, (2) and (8) hold.

To analyze the complexity of SMP we compare it with the minimum cover problem. Given a channel assignment problem, algorithm *SMP-to-MinCover* transforms this problem to a family of sets $\mathcal{F}$ such that any set $s \in \mathcal{F}$ is a transmission set. Then the objective of SMP can be seen to be the objective of finding a minimum cover for $\mathcal{F}$.

Algorithm 2 *SMP-to-MinCover*

Input: (G, c) of SAP
Output: $\mathcal{F}$, an instance of *MinCover* Problem

1: $H(V_a, E_a) \leftarrow SAP\text{-}to\text{-}MaxClique(G, c)$
2: $\mathcal{F} \leftarrow$ All complete subgraphs of H

The complexity of algorithm *SMP-to-MinCover* is equivalent to that of SAP with the same set of constraints. To solve SMP, we need to find a minimum cover of $\mathcal{F}$. The decision version of the minimum cover problem is to determine if a cover of size T or less exists given a family of sets $\mathcal{F}$, where the cardinality of every set in $\mathcal{F}$ is upper bounded by k.

Problem 4 Minimum k-cover problem.
Instance: Family $\mathcal{F}$ of subsets over E such that $\forall s \in \mathcal{F}$, $|s| \leq k$, and a positive integer T.
Question: Does $\mathcal{F}$ contain a cover for E of size T or less, that is, is there a subset $\mathcal{F}' \subseteq \mathcal{F}$ with $|\mathcal{F}'| \leq T$ and such that $\bigcup_{e \in \mathcal{F}'} e = E$?

This problem is proved to be NP complete in [15] by reducing it to the exact cover problem, which is a known NP complete problem. However, if the maximum cardinality of sets in $\mathcal{F}$ is two (i.e. $k=2$), then the problem of finding minimum cover can be solved in polynomial time.

5 Complexity of Channel Assignment Problems

In this section we analyze the complexity of the sub-channel assignment (further referred to as channel assignment) problems presented in Table 3. Note that all these problems belong to the class NP [15] in general. However, in the remaining of the section, for each of the channel assignment problems we either show that it can be solved in polynomial time or we prove that it is NP hard under certain conditions.

5.1 Unfair and Distance Constrained Channel Assignment

The unfair and distance constrained channel assignment problem can be stated as follows.

Problem 5 *unfair-δ-CAP.*
Given a channel assignment problem instance (G, c) with $c=(A_c, unfair, \delta_{th}, |E|)$ and k, a positive integer, is there a solution to *unfair-CAP* of size k i.e., is there a subset $E_t \subseteq E$ with $|E_t| \geq k$ such that all the constraints in c are satisfied?

To determine the complexity of solving this problem, we explore the implication of distance constraint on NP completeness of the unfair channel assignment problem. Our first step in this direction is to arrive at an upper bound on the number of simultaneous transmissions that are possible for a wireless network spread over a geographical area A with distance constraint δ_{th} and maximum transmission radius R_{max}. It can shown following the work in [5] that under the protocol model the number of simultaneous transmissions J is upper bounded by $\frac{4c}{\pi \delta_{th}^2 r_{max}^2}$, where $0 < c < 1$ is a suitable constant and $r_{max} = \frac{R_{max}}{\sqrt{A}}$ is the normalized maximum transmission radius.

Theorem 1 *For a given distance constraint δ_{th}, radius r_{max}, and area A of the wireless network, the unfair channel assignment problem can be solved in polynomial time, if $|E| \gg J$.*

   Proof. To show that this problem is polynomial time solvable, we propose an algorithm *unfair-δ-CAP* that solves every instance of the given problem and prove that the algorithm runs in polynomial time.

Algorithm 3 *unfair-δ-CAP*

Input: $(G, c), r_{\max}$.

Output: E_1, set of links assigned to the first sub-channel.

1:   $J = \frac{4c}{\pi \delta_{th}^2 r_{\max}^2}$

2:   **if** $|E| \gg J$ **then**

3:       $H(V_a, E_a) \leftarrow$ *SAP-to-MaxClique*(G, c)

4:       *MaxClique*$(V_{mc}, E_{mc}) \leftarrow$ *Naive-MC*(H, J)

5:       $E_1 \leftarrow$ *MaxClique*(V_{mc})

6:   **end if**

We know that *SAP-to-MaxClique*(G) runs in $\Theta(|E|^2)$. Note that $J \ll |E|$ here is a constant independent of E which implies that J remains constant with respect to an increase in the number of edges E. Using this fact we can compute the complexity of *Naive-MC* algorithm as

$$Naive\text{-}MC(G, J) \cong \Theta(\sum_{i=1}^{J} \binom{|E|}{i})$$

$$\cong \Theta(|E|^J), \forall J \ll |E|$$

The complexity of *unfair-δ-CAP* algorithm is $\mathcal{O}(|E|^J)$, which is polynomial in $|E|$.   □

5.2 1-Fair and Distance Constrained Channel Assignment

The *1-fair* distance constrained channel assignment is defined as follows:

Problem 6 *1-fair-δ-CAP.*

Instance: A channel assignment problem instance (G, c) with $c = \{A_c, \delta_{th}, |E|\}$ such that $|E| \gg J$ a positive integer T, a set of positive integers $\{k_1, \ldots, k_T\}$.

Question: Is there an 1-*fair-δ* constrained channel assignment for G with total number of sub-channels lesser or equal to T and each sub-channel of size k? That is, is there a channel assignment $\{E_t\}$, with $|\{E_t\}| \leq T$ and $\forall E_{t_i} \in \{E_t\}, |E_{t_i}| \geq k_i$ such that all the constraints in c are satisfied ?

In the following lemma we show a special class of *1-fair-δ-CAP* for which, every input instance of the *1-fair-δ-CAP* can be represented as a family $\mathcal{F}$

of feasible transmission sets with the cardinality of $\mathcal{F}$ polynomial in $|E|$. In the related corollary we show that the above transformation of *1-fair-δ-CAP* input instance to family $\mathcal{F}$ can be done in polynomial time.

Lemma 1 *Given an instance of 1-fair-δ-CAP in a wireless network with area A, maximum radius $r_{\max}$ and constraint δ_{th}, the number of feasible transmission sets is polynomial in $|E|$ if $|E| \gg J$.*

    Proof. Follows from Theorem 1.   □

Corollary 1 *Transformation $\pi : 1$-fair-δ-CAP $\to \mathcal{F}$ can be done in polynomial time for a given wireless network with area A, maximum radius $r_{\max}$ and constraint δ_{th} when $|E| \gg J$.*

    Proof. Follows from Theorem 1, using Naive-MC algorithm.   □

The optimal channel assignment $\{E_t\}$ of *1-fair-δ-CAP* is a collection of transmission sets. Since $\mathcal{F}$ is the set of all feasible transmission sets given the constraints $\mathfrak{c}$, we have $\{E_t\} \subseteq \mathcal{F}$.

Lemma 2 *If two sets $s_i, s_j \in \mathcal{F}$ s.t $s_i \subset s_j$ then $s_i \notin \{E_t\}$, where $\{E_t\}$ is the optimal 1-fair-δ-CAP solution.*

    Proof. Assume $s_i \in \{E_t\}$. This implies $s_j \in \{E_t\}$, since *1-fair-δ-CAP* outputs sets with maximum cardinality. Since $\{E_t\}$ is optimal, $s_i \notin \{E_t\}$, a contradiction.   □

We can hence represent the family $\mathcal{F}$ by its maximal feasible sets $\mathcal{F}^{\max}$ such that it is not possible to have two sets s_i, s_j such that $s_i \subset s_j$; and s_i, $s_j \in \mathcal{F}^{\max}$. Note that any optimal solution on $\mathcal{F}$ will be completely contained in $\mathcal{F}^{\max}$. However the use of $\mathcal{F}^{\max}$ instead of $\mathcal{F}$ may reduce the search complexity.

Theorem 2 *1-fair-δ-CAP is NP-complete.*

    Proof. The maximum cardinality of a transmission set in the family $\mathcal{F}^{\max}$ can be upper bounded by J (from Lemma 1). The objective of *1-fair-δ-CAP* will then be to pick a subset $f \subseteq \mathcal{F}$ such that every edge in E is covered by f. For $k = J$ and $\{E_t\} = f$, the well known k-Set Cover problem (see Definition4) and *1-fair-δ-CAP* are equivalent (from Lemma 1 and Corollary 1). Hence for $J \geq 3$ this problem is NP complete and for $J \leq 2$ optimal solutions can be found in polynomial time using matching techniques [8,9]. □

5.3 Fair Distance Constrained Channel Assignment

The fair distance constrained channel assignment can be defined as follows.

Problem 7 *fair-δ-CAP.*
Instance: A channel assignment problem instance (G, c) with $c = \{A_c, fair, \delta_{th}, |E|\}$ such that $|E| \gg J$ a positive integer T, a set of positive integers $K=\{k_1, \ldots, k_T\}$.
Question: Is there an *fair-δ-CAP* for G with number of sub-channels lesser or equal to T and each sub-channel of size K? That is, is there a channel assignment $\{E_t\}$, with $|\{E_t\}| \leq T$, $\forall E_{t_i}, E_{t_j} \in \{E_t\}$, $E_{t_i} \cap E_{t_j} = \phi$ with $|E_{t_i}| \geq k_i$ such that all other constraints in c are satisfied?

From Lemma 1 and Corollary 1 it follows that every input instance of *fair-δ-CAP* can be represented as a family $\mathcal{F}$ of feasible transmission sets in polynomial time. The objective of *fair-δ-CAP* is then to find a minimal subcover f which has no overlap. This minimal subcover f will form the optimal channel assignment $\{E_t\}$ satisfying all the constraints in c.

Theorem 3 *fair-δ-CAP is NP complete.*

Proof. For $k = J$ and $\{E_t\} = f$, the SET COVERING II problem [8] is equivalent to *fair-δ-CAP*. Hence, for $J \geq 3$ this problem is NP complete and for $J \leq 2$ optimal solutions can be found in polynomial time. □

5.4 Fair Distance Constrained and Balanced Channel Assignment

The *fair* distance constrained and balanced channel assignment is defined as follows:

Problem 8 *fair-δ-Balanced-CAP.*
Instance: A channel assignment problem instances (G, c) with $c = \{A_c, fair, \delta_{th}, \epsilon_{th}\}$ such that $|E| \gg J$ a positive integer T, a set of positive integers $\{k_1, \ldots, k_T\}$.
Question: Is there *fair-δ-Balanced-CAP* for G with number of sub-channels lesser or equal to T and each sub-channel of size k? That is, is there a channel assignment $\{E_t\}$, with $|\{E_t\}| \leq T$ and $\forall E_{t_i}, E_{t_j} \in \{E_t\}$, $E_{t_i} \cap E_{t_j} = \phi$ with $|E_{t_i}| \geq k_i$ and ϵ_{th} balanced such that all other constraints in c are satisfied?

Theorem 4 *fair-δ-Balanced-CAP is NP complete.*

Proof. Restrict to partition problem [15] by having $T = 2$, $\epsilon_{th} = 0$ and $\forall e_i$, $s(e_i) = 1$. □

6 Algorithms to Solve Distance Constrained Unfair, 1-fair and fair CAP

We have seen that algorithm *SMP-to-MinCover* transforms any given instance of channel assignment problem into a set $\mathcal{F}$ of subsets $\{E_t\}$ of the edges E of the wireless network. In this section we propose another algorithm called *Gen-$\mathcal{F}^{\max}$* which is similar to *SMP-to-MinCover* but with a huge reduction in complexity. Here, instead of generating the entire family $\mathcal{F}$, we generate only the maximal family $\mathcal{F}^{\max}$ using a novel tree pruning approach. For any given CAP instance, *Gen-$\mathcal{F}^{\max}$* constructs the corresponding maximal family $\mathcal{F}^{\max}$.

Algorithm 4 *Gen-$\mathcal{F}^{\max}$*

1: $i = 1$
2: $F_i = \{\{1\}\{2\} \ldots \{|E|\}\}$
3: $F_{i+1} = Extend\text{-}Family(F_i, (C))$
4: **while** $F_{i+1} \neq \phi$ **do**
5:     $\forall f_i \in F_i; F_i^{\max} = \{f_i : f_i \not\subseteq \bigcup_{k=1}^{|F_{i+1}|} F_{i+1}(k)\}$
6:     $i = i + 1$
7:     $F_{i+1} = Extend\text{-}Family(F_i, C)$
8: **end while**
9: $F^{\max} = \bigcup_i F_i^{\max}$

procedure $F_{ext} = Extend\text{-}Family(F_{orig}, C)$

1: $F_{ext} = \{\phi\}$
2: **for** $i = 1 : length(F_{orig})$ **do**
3:     **for** $j = F_{orig}(i, |F_{orig}(i)|) + 1 : |E|$ **do**
4:         **if** $Satisfy(\{F_{orig}(i)j\}, C)$ **then**
5:             $F_{ext} = F_{ext} \cup \{F_{orig}(i)j\}$
6:         **end if**
7:     **end for**
8: **end for**

An instance of *Gen-$\mathcal{F}^{\max}$* on a six edge example network (Figure 4) is shown in Figure 5. Initially every edge in the family is allocated to a singleton set. Thus the initial family comprises of as many subsets as the number of edges. For the given example, the initial family is $\{$ $\{1\}$ $\{2\}$ $\{3\}$ $\{4\}$ $\{5\}$ $\{6\}$ $\{7\}$ $\}$. In the next and the following rounds two fundamental operations take place. The first operation is expansion, where each of the subsets in the family

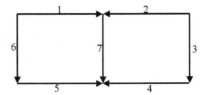

Figure 4 Example network with six edges.

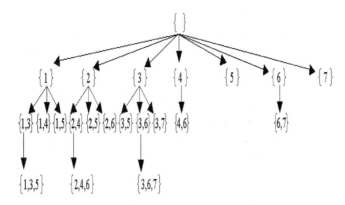

Figure 5 Instance of *Gen-$\mathcal{F}^{\max}$* on a six edge example network.

is expanded by including one more unique edge. For the given example, the family obtained after first expansion is { {1, 3} {1, 4} {1, 5} {2, 4} {2, 5} {2, 6} {3, 5} {3, 6} {3, 7} {4, 6} {6, 7} {1} {2} {3} {4} {5} {6} {7} }. Here, we do not see {1, 2},{1, 6},{2, 3}..etc as they do not satisfy the constraint set. The second operation is pruning, where the subsets which are in-turn subsets of larger sets are eliminated from the family. For the given example, the family obtained after first pruning is { {1, 3} {1, 4} {1, 5} {2, 4} {2, 5} {2, 6} {3, 5} {3, 6} {3, 7} {4, 6} {6, 7} }. These two operations are performed iteratively until expansion stops. For the given example, the final family is { {1, 3, 5} {2, 4, 6} {3, 6, 7} }.

Theorem 5 *For any given instance of CAP, Algorithm Gen-$\mathcal{F}^{\max}$ constructs a maximal family $F^{\max}$.*

Proof. Follows by induction on $\bigcup_1^i \mathcal{F}_i$. □

The following lemma and theorem establish that the proposed pruning algorithm has polynomial order complexity.

Lemma 3 *The probability α of co-existence of a pair of links is $1 - \pi \delta_{th}^2 r_{max}^2$.*

Proof. The probability of co-existence of any pair of links in the network is given by the probability that the receiving node of the second link is at least $\delta_{th} r_{max}$ apart from the receiving node of the first link (see Figure 3). That is, the probability the second receiving node being outside the circle with radius $\delta_{th} r_{max}$ centered around the first receiving node. Clearly, this probability is equal to $\alpha = 1 - \pi \delta_{th}^2 r_{max}^2$. □

Theorem 6 *The expected number of feasible transmission sets is upper bounded by $(|E| - \pi \delta^2 R_{max}^2 n_o)^J$, where $J = \frac{4c}{\pi \delta_{th}^2 r_{max}^2}$, $r_{max} = R_{max}/\sqrt{A}$ and n_o is the edge density in the network.*

Proof. The expected number of pairs of links in the network is $\alpha\binom{n}{2}$, where $n = |E|$. For any transmission set of cardinality k with $k > 1$, the probability of coexistence is $\alpha^{\binom{k}{2}}$. Therefore, the expected number of feasible transmission sets is given by

$$E(|\mathcal{F}|) = \sum_{k=2}^{J} \alpha^{\binom{k}{2}} \binom{n}{k} \tag{10}$$

$$= \sum_{k=2}^{J} \alpha^{\frac{k(k-1)}{2}} \frac{n!}{(n-k)!k!} \tag{11}$$

$$\leq \sum_{k=2}^{J} \alpha^{k^2} n^k \tag{12}$$

$$\leq \sum_{k=0}^{J} (\alpha n)^k \tag{13}$$

if $(\alpha n) > 1$, $E(|\mathcal{F}|) \cong \mathcal{O}(\frac{(\alpha n)^{J+1}-1}{(\alpha n)-1})$
if $(\alpha n) < 1$, $E(|\mathcal{F}|) \cong \mathcal{O}(\frac{1}{1-(\alpha n)})$
if $(\alpha n) = 1$, $E(|\mathcal{F}|) \cong \mathcal{O}(J)$.

The edge density of the network is given by $n_o = |E|/A$. Hence from the above three cases we see that the upper bound of $E(|\mathcal{F}|)$ is $\mathcal{O}(\frac{(\alpha n)^{J+1}-1}{(\alpha n)-1}) \cong \mathcal{O}(\alpha n)^J = \mathcal{O}(|E| - \pi \delta^2 R_{max}^2 n_o)^J$. □

Since *Gen-$\mathcal{F}^{max}$* looks only at the feasible transmission sets, the running time of *Gen-$\mathcal{F}^{max}$* is $\mathcal{O}(E(|\mathcal{F}|))$.

6.1 Optimal *unfair-CAA*

An optimal unfair channel assignment algorithm is the following. Choose the set with the largest cardinality in $\mathcal{F}^{\text{max}}$ generated by $Gen\text{-}\mathcal{F}^{\text{max}}$. The solution is optimal because, $Gen\text{-}\mathcal{F}^{\text{max}}$ generates the family of all possible large transmission sets. That is, there cannot be a transmission that is larger than the all of the transmission sets in $\mathcal{F}^{\text{max}}$ and satisfy all the specified constraints.

When $|E| \gg J$ this algorithm produces optimal solution for *unfair-CAP* and the running time is same as the $Gen\text{-}\mathcal{F}^{\text{max}}$ algorithm.

6.2 Approximate *1-fair-CAA*

We know that there can be no deterministic algorithm that solves *1-fair-CAP* in polynomial time. Heuristic algorithms [8] that solve minimum cover problem can be used on *1-fair-CAP*. The input instances of heuristic algorithms that solve minimum cover problem should be mapped from instances of $\{\mathcal{F}^{\text{max}}, E\}$ of *1-fair-CAP* and the minimum cover output instances should be mapped to channel assignment of *1-fair-CAP*.

We use the greedy approximation algorithm discussed in [3]. This algorithm works in a greedy fashion selecting the set that covers the maximum number of uncovered elements in every iteration. The run time complexity of this algorithm is $\mathcal{O}(\sum_{s \in \mathcal{F}^{\text{max}}} |s|)$. It is shown that this algorithm returns a set cover that has an approximation ratio bound of $H(\max\{|s| : s \in \mathcal{F}^{\text{max}}\})$ over the optimal solution, here $H(d)$ denotes the dth harmonic number $H(d) = \sum_{i=1}^{d} 1/i$. From Section 5.1 we know that $\max\{|s| : s \in \mathcal{F}^{\text{max}}\} \leq J$ where J is independent of E. Hence the approximation ratio of greedy set cover for this case is $H(J)$.

6.3 Approximate *fair-CAA*

By Theorem 5.3 fair distance constrained channel assignment problem is NP complete. The greedy heuristic based on graph coloring [13] can be used directly for the $\delta_{th} = 0$ case. However, for distance constrained assignment a slight modification to the unified algorithm given in [13] generates fair channel assignment satisfying the distance constraints.

7 Simulation Results

7.1 Performance Measures

To evaluate the effectiveness of the channel assignment algorithms proposed in Section 6, we use the following performance measures.

7.1.1 Number of Sub-Channels

The total number of sub-channels (denoted by T) used by the assignment technique is an important performance metric. In any protocol, there is some overhead associated with each sub-channel and this overhead adds up for every new sub-channel. When the sub-channels are in the form of time slots, then more number of time slots directly translates to increased delay. Hence it is generally preferable to have a schedule that uses least number of sub-channels while maintaining high link capacities.

7.1.2 Capacity

We use two different notions of capacity. The first one is *protocol capacity*, which is the average number of links scheduled to transmit per sub-channel. That is,

$$C = \frac{1}{T} \sum_{t=1}^{T} |E_t| \qquad (14)$$

The second notion for capacity is the *sum rate capacity*, which is the sum of Shannon's capacity [11] for each link divided by the total number of sub-channels used.

7.1.3 Imbalance

Imbalance is the maximum absolute difference in the number of links between two sub-channels.

7.2 Performance of Proposed Channel Assignment Algorithms

We evaluate the performance of the *unfair*, *1-fair* and *fair* channel assignment algorithms for wireless networks with varying node densities. We use both the grid network model and the random network model. The area of the network used in the simulation is 100×100 square meters.

For the grid networks, we fix 16 nodes on the network. Initially we do not assume any distance ratio constraint. We then evaluate both the protocol and the sum rate capacity for Unfair and 1-fair channel assignment. The same

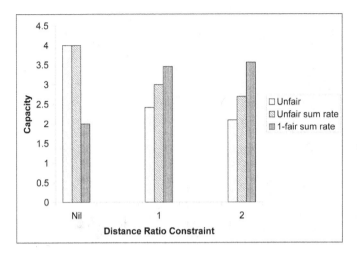

Figure 6 Capacity vs. distance ratio constraint for a grid network with area of 100 × 100 square meters.

experiment is performed for distance ratio constraints with δ equal to 1 and 2. The corresponding results are plotted in Figure 6. Note that Nil represents the absence of distance ratio constraint. We can observe from the figure that as the distance ratio constraint increases the capacity defined by protocol model seems to decrease where as the sum rate capacity, which is the true capacity of the network increases. We can further observe that there is a gain of about 70% in sum rate capacity in 1-fair channel assignment when distance ratio constraint is introduced. This suggests that SINR/distance constrained protocols result in better channel assignments in terms of sum rate capacity.

For the random network model, we fixed the distance ratio constraint δ_{th} to 3. The maximum transmission range of the network is fixed to 10 meters. Figure 7 shows the comparison of the capacities of unfair, 1-fair and fair channel assignment algorithms as a function of average number of edges in the network. We can see that as the fairness constraint is relaxed from exact fairness to unfair the capacity achieved increases. For example, Figure 7 shows that for a network with 80 edges, there is a 100% gain in capacity when the fairness is relaxed from exact fairness to 1-fairness and 25% gain when fairness constraint is relaxed from 1-fairness to unfair. The number of sub-channels used to cover the network by 1-fair and fair are given in Figure 8. We can observe that the heuristic based on covering and coloring result in

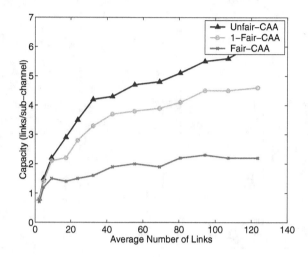

Figure 7 Performance of unfair, 1-fair, and fair CAA in terms of average number of links per sub-channel versus the number of edges, in a network with area 100×100 square meters, distance constraint $\delta_{th} = 3$ and maximum range = 10 meters.

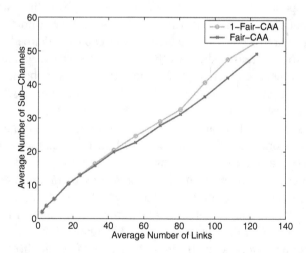

Figure 8 Performance of 1-fair and fair CAA in terms of the number of sub-channels used versus the number of edges, in a network with area 100×100 square meters, distance constraint $\delta_{th} = 3$ and maximum range is 10 meters.

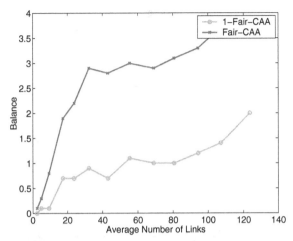

Figure 9 Performance of 1-fair and fair CAA in terms of balance versus the number of edges, in a network with area 100×100 square meters, distance constraint $\delta_{th} = 3$ and maximum range is 10 meters.

comparable performance over all networks considered. The imbalance of 1-fair and fair channel assignments over different sub-channels is given in Figure 9. Here we can see that the maximum value of imbalance for *1-fair CAA* is equal to 2. This indicates that all the sub-channels have almost the same number of edges.

To study the impact of distance constraint δ_{th} on the capacity of the channel assignment, we fix the number of edges in the network to 24 and vary the distance constraint from 0 to 10. Figure 10 plots the reduction in capacity due to increase in the distance constraint (δ_{th}). Here we can observe that for smaller distance constraints 1-fair and unfair CAA perform comparably, however as the distance constraint is increased 1-fair CAA performs poorly compared to unfair CAA. We can also observe that exact fair channel assignment quickly converges to the one link per sub-channel worst case behavior compared to 1-fair and unfair assignments.

We also study the capacity of the assignment for networks with different maximum transmission range, R_{max}. We fix $\delta_{th} = 3$, number of nodes to 30 and vary the maximum allowable transmission range from 1 to 20 meters. Figure 11 plots the capacity versus increase in the maximum allowable range. We can observe that the capacities of channel assignments increase up to a point and then decreases. This is because at lower transmission range, the network has very few edges and many disconnected nodes. As the transmission

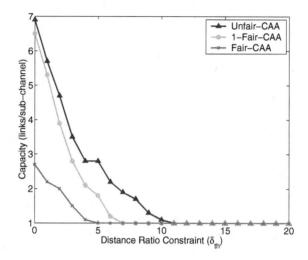

Figure 10 Performance of unfair, 1-fair, and fair CAA in terms of average number of links per sub-channel versus the distance constraint δ_{th}, in a network with 24 edges, area 100×100 square meters and maximum range is 10 meters.

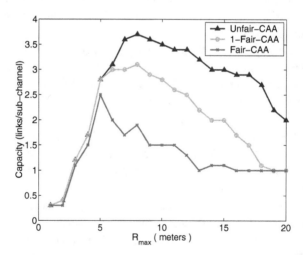

Figure 11 Performance of unfair, 1-fair, and fair CAA in terms of average number of links per sub-channel versus the maximum range, in a network with 30 nodes, area 100×100 square meters and $\delta_{th} = 3$.

range increases the number of edges and hence the capacity also increases and reaches the optimum. The decrease in capacity can be explained by the bound J where r_{max} appears in the denominator.

8 Conclusion

In this paper we systematically studied the channel assignment problem in presence of fairness, SINR and balance constraints. Both grid and random models of wireless network were used. Several channel assignment problems based on specific constraints were identified. We showed that a class of unfair channel assignment problems can be optimally solved in polynomial time. We also showed that with fair and 1-fair constraints the channel assignment problem remains NP complete. Optimal unfair channel assignment algorithm, approximate 1-fair channel assignment algorithm and approximate fair channel assignment algorithm were proposed. Simulation results revealed that there is a gain in capacity of 100% when fairness constraint is relaxed from exact fairness to 1-fairness and 25% when relaxed from 1-fairness to unfair. It was observed that for smaller values of the distance constraint 1-fair and unfair perform comparably. We also saw that 1-fair channel assignment was more balanced than fair channel assignment.

Acknowledgement

This work was supported by NSF #0916180 and NSF #0917008.

References

[1] E. Arikan. Some complexity results about packet radio networks. IEEE Transactions on Information Theory, 30(4):681–685, July 1984.

[2] I. Chlamtac and A. Lerner. Fair algorithms for maximal link activation in multihop radio networks. IEEE Transactions on Communications, COM-35(7):739–746, July 1987.

[3] T. Cormen, C. Leiserson, R. Rivest, and C. Stein. Introduction to Algorithms. McGraw-Hill, second edition, 2002.

[4] A. Ephremides and T. Truong. Scheduling broadcasts in multihop radio networks. IEEE Transactions on Communications, 38(4):456–460, April 1990.

[5] P. Gupta and P. Kumar. The capacity of wireless networks. IEEE Transactions on Information Theory, 46(2):388–404, March 2000.

[6] B. Hajek and G. Sasaki. Link scheduling in polynomial time. IEEE Transactions on Information Theory, 34(5):910–917, September 1988.

[7] J. L. Hammond and H. B. Russell. Properties of a transmission assignment algorithm for multi-hop packet radio networks. IEEE Transactions on Wireless Communications, 3(4):1048–1052, July 2004.

[8] D.S. Johnson. Approximation algorithms for combinatorial problems. In STOC'73: Proceedings of the Fifth Annual ACM Symposium on Theory of Computing, pages 38–49. ACM Press, New York, 1973.

[9] R. Karp. Reducibility among combinatorial problems. In R.E. Miller and J.W. Thatcher (Eds.), Compexity of Computer Computations, pages 85–103. Plenum Press, New York, 1972.

[10] S. Kutten and I. Chlamtac. A spatial reuse tdma.fddma for mobile multi-hop radio networks. In INFOCOM, 1985.

[11] A. Leon-Garcia and I. Widjaja. Communication Networks: Fundamental Concepts and Key Architectures. McGraw-Hill School Education Group, 1999.

[12] C.G. Prohazka. Decoupling link scheduling constraints in multi-hop packet radio networks. IEEE Trans. Comput., 38(3):455–458, 1989.

[13] S. Ramanathan. A unified framework and algorithm for channel assignment in wireless networks. Wireless Networks, 5:81–94, 1999.

[14] R. Ramaswami and K. Parhi. Distributed scheduling for broadcasts in a radio network. In INFOCOM, 1989.

[15] M.R.Garey and D.S. Johnson. Computers and Intractability, A Guide to the Theory of NP-Completeness. Freeman and Company, New York, 2003.

[16] P. Värbrand, D. Yuan, and P. Bjorklund. Resource optimization of spatial TDMA in ad hoc radio networks: A column generation approach. In INFOCOM, 2003.

Biographies

Chetan Mathur received his Ph.D. in Computer Engineering at Stevens Institute of Technology, New Jersey, USA in 2007. He has an MS in Computer Engineering from Stevens Institute of Technology, New Jersey, USA in 2003. Part of Chetan's MS thesis was patented by Stevens Institute of Technology. He was born in Bangalore, India in 1981. He received his BE degree in Computer Science from Visveshwaraiah Institute of Technology, Bangalore, India in 2002. Chetan has published several research papers in the fields of Cryptography, Coding theory and Dynamic spectrum access. He has also received numerous awards including the IEEE best student paper award presented at IEEE Consumer Communications and Networking Conference (CCNC 2006) and the IEEE student travel grant award presented at International Conference on Communications (ICC 2005). He is currently employed in the financial industry.

M.A. Haleem graduated from Stevens Institute of Technology with a Ph.D. in Electrical and Computer Engineering. His research interests are in the areas of wireless communications and signal processing. He is currently an Assistant Professor at KFUPM, Saudi Arabia.

K.P. (Suba) Subbalakshmi is an Associate Professor at Stevens Institute of Technology. Her research interests are in the areas of cognitive radio networks, wireless network security, media forensics as well as social networks. She is the Vice-Chair North America region of IEEE Technical Committee on Cognitive Networks. She has given several key-note addresses, plenary talks and tutorials on DSA security at several international conferences. She has also served as a panelist on cognitive radio network security at international conferences including IEEE Dynamic Spectrum Access: Collaboration between the Technical, Regulatory and Business Communities, IEEE ICC, IEEE Sarnoff Symposium etc. She was a Guest Editor of the *EURASIP Journal on Advances in Signal Processing*, Special Issue on Dynamic Spectrum Access for Wireless Networks. Her work is/has been supported by the National Science Foundation, National Institute of Justice, DoD agencies as well as the Industry. Suba is also the co-founder of two companies that seek to commercialize some of her research work. One of these is Dynamic Spectrum LLC which commercializes her work in Dynamic Spectrum Access Networks.

R. Chandramouli (Mouli) is the Thomas Hattrick Chair Professor of Information Systems in the Department of Electrical and Computer Engineering (ECE) at Stevens Institute of Technology and Co-Director of the Information Networks and Security (iFINITY) research laboratory. He is a Co-Founder of Dynamic Spectrum, LLC – a startup offering cloud-enabled cognitive radio technologies for various markets including consumer communications, public safety, and the DoD; and Jaasuz.com that provides a suite of advanced text forensics technologies to verify trust in documents. His research spans the areas of wireless networking, social media analytics/security and computational psycho-linguistic text mining.

Authenticated Encryption for Low-Power Reconfigurable Wireless Devices

Samant Khajuria and Birger Andersen

Aalborg University, Denmark, and Copenhagen University College of Engineering, Lautrupvang 15, 2750 Ballerup, Denmark; e-mail: skh@es.aau.dk, bia@ihk.dk

Abstract

With the rapid growth of new wireless communication standards, a solution that is capable of providing a seamless shift between existing wireless protocols and high flexibility as well as capability is crucial. Technology based on reconfigurable devices offers this flexibility. In order to avail this enabling technology, these radios have to propose cryptographic services such as confidentiality, integrity and authentication. Therefore, integration of security services to these low-power devices is very challenging and crucial as they have limited resources and computational capabilities.

In this paper, we present a crypto solution for reconfigurable devices. The solution is a single pass Authenticated Encryption (AE) scheme that is designed for protecting both message confidentiality and its authenticity. This makes AE very attractive for low-cost low-power hardware implementation. For test and performance evaluation the design has been implemented in Xilinx Spartan-3 sxc3s700an FPGA. Additionally, this paper analyzes different hardware architectures and explores area/delay tradeoffs in the implementation.

Keywords: authenticated encryption, confidentiality, message authentication, FPGA, wireless devices.

Journal of Cyber Security and Mobility, Vol. 1, 189–203.

1 Introduction

Over the past decade, wireless devices have become an indispensable part of our life. With time like every other technological device, the features and capabilities of the wireless devices are also evolved. Nowadays, devices like mobile phones are able to do lot more in addition to their traditional roles of voice communication. This has motivated new application domains for wireless networks. For example, wireless sensor networks (WSNs) are used in various applications, including environmental monitoring, military systems, health care, etc. Vehicular ad hoc networks (VANETs) promise road safety, while disruption-tolerant networks (DTNs) bring low-cost best-effort connectivity to challenged environments with little or no infrastructure [1]. Furthermore, the concept of Internet of Things (IoT) has picked up surge of interest with enormous applications in home and industry. Due to the advancement in the field, these networks offer a world of truly ubiquitous computing. With these additional abilities of the radios that are applicable across a wide range of areas within the wireless infrastructure, these radios have to implement cryptographic services such as confidentiality, integrity and authentication.

Typically, devices are equipped with an antenna from where they receive the data and-then-they process and transmit. Since the devices are compact and wireless, they are highly energy constraint. Data processing and wireless communication count for the greatest part of the energy consumed by a device. Especially in case of sensors, the need to operate for longer period of time demands for better and careful management of power resources. On top of this security is very challenging and crucial as devices have limited resources and computational capabilities. In order to provide data confidentiality and other cryptographic services, there is a need for lightweight schemes that can promise similar security as compared to traditional cryptographic schemes.

Future visions of wireless devices are foreseen as the devices connecting to a wide range of different networks or devices. This can be achieved by changing the characteristics of the devices by making software changes. By doing this the devices can adapt to the user preferences and the operating environment and support multiple standards without requiring separate radios for each standard. The possibility of dynamically adapting according to the environment is through the re-configuration of device's components. More specifically, the re-configurability is the ability of adjusting operational parameters for the transmission on-the-fly without any modifications on the hardware components. Unlike implementing these functional blocks on inflexible Application Specific Integrated Circuits (ASICs) in the past, the technologies such as Field programmable Gate Arrays (FPGAs) are used to build radio functional blocks. FPGAs have reconfigurable capability and deliver flexibility of programmable architectures with power efficiency and performance. The reprogrammable nature of FPGAs makes them ideal for wireless devices, so any upgrades or changes in the operational parameters can be easily uploaded to the device without any hardware reconfigurations. FPGAs also allow the feature of partially reconfiguring the devices, the model is known as shared resource model. As compared to dedicated resource model, shared resources are capable of supporting ex., multiple waveforms across a single set of processing resources; this allows for much more efficient usage of the resources. Partial reconfiguration allows the replacement of one or multiple functional blocks with a different implementation while other portions are either being used by other applications or going unused. Without partial reconfiguration, it would be necessary to reconfigure entire FPGA. Using partially reconfigurable platform FGPAs for wireless devices will substantially decrease the component count of the devices and reduce power consumption while still providing the necessary functionality.

In this paper, we present a crypto solution for reconfigurable wireless devices. Section 2 summarizes the security issues and two main security objectives for wireless devices. Section 3 provides a brief overview of single pass authenticated encryption scheme. Section 4 details the architecture and overall design of the implementation, while the results are presented in Section 5. Finally conclusions are drawn in Section 6.

2 Security Objectives

In order to communicate between two or more devices or to enjoy the flexibility of reconfigurable radios to upgrade or adapt to user preferences many security countermeasures needs to be taken into account. Reconfiguring the radios has many benefits; however the ability to reconfigure radio functionalities with software may lead to many security problems such as unauthorized use of application and network services, unauthorized modification of software and manipulation of devices. For example, malicious software can be uploaded into the device that changes its radio frequency so that the device will no longer function within the regulated constraints. This could lead to the Denial of Service (DoS) attacks. Additionally, transmission of unencrypted data over insecure channel could compromise the confidentiality and integrity of the data.

The above mentioned security issues often concerns with two main security objectives: confidentiality and authenticity of the data. The objective of confidentiality is to keep the contents of the information secure and no one but the sender and authorized receivers are able to read the data. Authentication of message data verifies the origin and improper or unauthorized modification of data. In the past, confidentiality of the data was the main issue considered. This was mainly because no other security objectives such as authentication or integrity prevented to have access to the information. Only message encryption can protect data from eavesdroppers. However encryption of messages provides some sort of authentication but as compared to present authentication techniques it is weak and cannot be relied upon. In addition to confidentiality, authentication services have been implemented but as add on feature to provide extra information security. Encryption algorithms are used to ensure confidentiality while Message Authentication Codes (MAC) can be used to provide authentication. In past few years, techniques have been invented which can combine encryption and authentication into a single algorithm [2–4]. Combining these two security features and performing single pass operation we expect this will provide the following advantages for hardware implementation:

- The rapid growth of portable low-cost devices with limited area has opened a vast scope for compact circuit design opportunities. Implementation of a single algorithm instead of two separate algorithms definitely has less area requirements. Reduction in area requirements on chip is directly proportional to the reduction in cost.
- Small and compact designs tend to consume less power as compared to bulky designs. This is an attractive feature for low-power devices like Cellular phone, PDAs, smartcards and especially wireless sensor devices.
- Even though separate keys are used for encryption and authentication for better security of the system, both the keys are usually derived from the same master key. This will have a slight advantage with regards to the key storage issues over separate algorithms.
- Most of the new designs target performance goals like throughput and throughput-area trade-off. In many cases, combined schemes are based on block ciphers, and designers have tried to be efficient with the number of block cipher calls required for getting both confidentiality and

authentication from the algorithm. Based on the mode of the operations some of these combined schemes can run in parallel and achieve much higher speed than older techniques.

3 Authenticated Encryption

The cryptographic schemes that provide both confidentiality and authentication are called authenticated encryption schemes. The scheme is designed in such a way that the sender produces the ciphertext as well as an authentication tag which is verified by the receiver.

The authenticated encryption scheme consists of three algorithms: a key generation algorithm, an encryption algorithm and a decryption algorithm. The encryption algorithm takes a key, a plaintext and an initialization vector and it returns a ciphertext. Given the ciphertext and the secret key, the decryption algorithm returns plaintext when the ciphertext is authentic and invalid when the ciphertext is not authentic. The scheme is secure if it is both un-forgeable and secure encryption scheme [5]. When an attacker is not able to successfully produce a ciphertext C, a nonce N, and a tag σ (three parameters which maintain the integrity of the message) even if the attacker convinces the receiver to will believe that the sender was the originator, then the scheme is *un-forgeable*. The term *secure* is related towards confidentiality of the scheme, where confidentiality means, that an attacker cannot understand the contents of the message M, even after knowing the ciphertext C and the nonce N. One way to achieve this is to make the encryption scheme indistinguishable from a random permutation; this is a standard definition that is used in many security proofs such as the security proofs of the modes of operation for block ciphers.

The goal of authenticated encryption is to provide privacy and integrity. Two possible notations are used for the authenticity of AE, INT-PTXT (Integrity of the plaintexts) – $M = D_K(C)$ was never encrypted by the sender, it is computationally infeasible to produce a ciphertext decrypting to a message that is never encrypted by the sender and INT-CTXT (Integrity of the ciphertexts) – C was never transmitted by the sender, it is computationally infeasible to produce a ciphertext not previously produced by a sender. Privacy goals for encryption schemes consists of indistinguishability (advantage of a reasonable adversary determining what message was sent, M or M') and non-malleability (advantage of a reasonable adversary being able to change the message to be meaningful), each of which are considered under either chosen-plaintext or chosen-ciphertext attack. This leads to two indistinguishability notations of security IND-CPA (indistinguishability under a chosen plaintext attack), IND-CCA (indistinguishability under a chosen ciphertext attack) and two non-malleability security notations, namely NM-CPA (non-malleability under a chosen plaintext attacks), NM-CCA (non-malleability under chosen ciphertext attack).

3.1 ASC-1: An Authenticated Encryption Stream Cipher

The idea behind single pass Authenticated Encryption is to achieve faster encryption and message authentication by performing both the encryption and message authentication in a single pass as opposed to the traditional approach which requires two passes, i.e., one for encryption and other for authentication. In the past several single pass provable secure AE schemes have been proposed, for example, IACBC and IAPM [2]. Other provably secure AE schemes that use a block cipher as a building block were also presented in [3, 4]. In this section we describe a single pass authenticated encryption scheme ASC-1 [6]. The design of ASC-1 authenticated encryption scheme uses a four round Advanced Encryption Standard (AES) as a building block. The scheme uses single cryptographic primitive to achieve both message secrecy and authenticity. It is also shown that ASC-1 is secure if one cannot tell apart the case when the scheme uses random round keys from the case when the round keys are derived by a key scheduling algorithm.

As shown in Figure 1, ASC-1 is a single pass AE scheme that uses four round AES with 128-bit key as an underlying block cipher. ASC-1 is divided into two steps – Initial phase generation, Encryption in CFB (Cipher feedback)-like mode and authentication of the data. At the decryption side, same steps are repeated and the computed tag is matched with the received tag for verification.

Initial phase generation – Initial phase consists of an initialization vector X_0 and three keys $K_{1,0}$, $K_{2,0}$, $K_{3,0}$. To calculate these values ASC-1 uses 56-bit of the counter and applies 128-bit AES block cipher to $0^{70}\|00\|Cntr$, $0^{70}\|01\|Cntr$, $0^{70}\|10\|Cntr$, $l(M)\|00000011\|Cntr$, using Master key K_M, where $l(M)$ is the 64-bit representation of the bit length of the Message M.

Encryption Process – Before initializing encryption process, Keys $K_{1,0}$ and $K_{2,0}$ are concatenated together and AES-256 key scheduling algorithm is applied to derive 14 round keys. Keys K_2, K_3, K_4 and K_5 are used as round keys in the first round and Keys K_7, K_8, K_9 and K_{10} are used in the second round. Keys K_{11} and K_{12} are used as whitening keys in the first and second rounds of 4R-AES transformation respectively. In AES key scheduling round

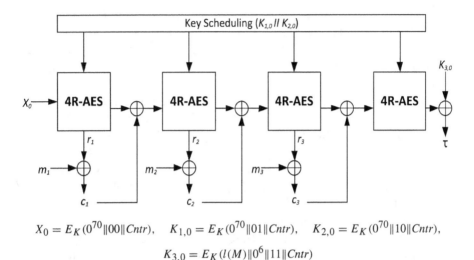

$$X_0 = E_K(0^{70}\|00\|Cntr), \quad K_{1,0} = E_K(0^{70}\|01\|Cntr), \quad K_{2,0} = E_K(0^{70}\|10\|Cntr),$$

$$K_{3,0} = E_K(l(M)\|0^6\|11\|Cntr)$$

Figure 1 The encryption algorithm of ASC-1. The message consists of three blocks. The ciphertext consists of the counter value, three ciphertext block and authentication tag.

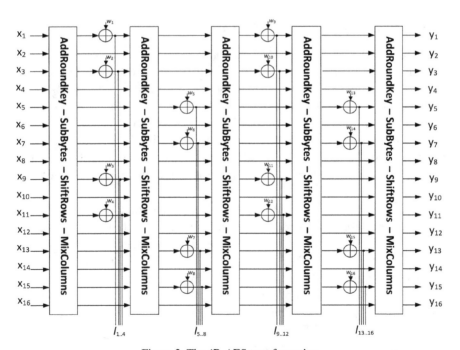

Figure 2 The 4R-AES transformation.

keys can either be generated on-the-fly or they can be stored in the internal memory. On the other hand, in ASC-1, because of using K_1 and K_{11} for key whitening, it is only possible to store the keys in the memory during the key setup phase, and then read them from this memory whenever they are required by the encryption/decryption unit.

ASC-1 Encryption Block consists of four round AES as shown in Figure 2. To initialize the encryption module, a 128 bit initialization vector is provided as an input to the ASC-1 encryption algorithm. ASC-1 performs a number of transformations to the input data to give a 128-bit leak $l_1, l_2, \ldots, l_{16}$ and output state $y_1, y_2, \ldots, y_{16}$. ASC-1 stream cipher performs four discrete transformations: *AddRoundKey*, *SubBytes*, *ShiftRows* and *MixColumns*. Four bytes are leaked at the end of every round and positions of the leaks depend on the number of the round (even or odd). Finally, a whitening key byte is added before each extracted byte. The AES-256 key scheduling algorithm is again applied to $K_{13} \| K_{14}$ to derive 14 keys that are used by the third and the fourth 4R-AES transformation, and the process is repeated as long as we need new keys.

4 Proposed ASC-1 Architecture

The high-level architectural organization of the ASC-1 encryption core is presented in Figure 3. The system is divided into five logical blocks. The initial input interface is responsible for feeding data to the key logic and the processing core. Key logic handles all the key scheduling operations and processing core block performs all the main encryption process. SBox block is a ROM that is used for the SubBytes transformation by key logic and core block. Finally the control unit is used for the synchronization and communication with the external logic. Let us further look into the functionality of each logic block in detail.

Initial input interface – For initial phase generation, i.e., initialization vector X_0 and three keys $K_{1,0}$, $K_{2,0}$, $K_{3,0}$, a new counter/nonce is loaded. The initial input interface concatenates the values of the counter with the predefined values stored in the local registers. The processing core unit is then notified that an initial state is available for processing.

Key Logic – In the above mentioned scheme, every encryption round requires a new round key. Once the new key is loaded, the key logic block starts generating round keys based on a single external key. Three possible approaches can be used to generate round keys – Online approach, Offline or stored-key approach and use of an external source ex., key generator or an

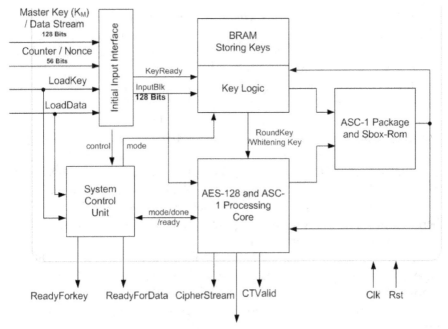

Figure 3 Block diagram of ASC-1.

external processor. Our design is based on "offline" or "stored-key" approach, where all the round keys are calculated upon the reception of the initial cipher key before the start of encryption and stores them in a local memory. The memory is accessed at every encryption round in order to provide the necessary round key. Opting for stored-key approach has many advantages in our design as compared to "online" or external source approach ex., for initial phase generation same key (K_M) is used to encrypt initialization vector (IV), two initial keys ($K_{1,0}$, $K_{2,0}$) for key scheduling for ASC-1 encryption and key ($K_{3,0}$) for authentication of data. The round keys derived from the Master key (K_M) is stored in the memory and during the encryption process right round key is accessed from the memory to perform encryption operation. In case encryption of stream data, 14 round keys are derived by loading 256-bit key, i.e., $K_{1,0} \| K_{2,0}$ to the key logic unit for key expansion. The key logic block performs two main functions. The key expansion process and read/write round keys to the memory block. The first one is performed whenever a new cipher key is inserted to the block and second one is to fetch round keys from the local memory for encryption process.

AES and ASC-1 processing core – The processing core block consists of AES-128 and ASC-1 encryption process. AES encryption core is used only for the generation of Initialization vector and keys used in ASC-1. Once the IV and the keys are encrypted using AES-128, keys are fed into the key logic block for the calculation of round keys and IV is used to initiate ASC-1 scheme. The underlying block used in ASC-1 is AES, so same transformations are applied to the block but in different order. AddRoundKey transformation is the first block and after MixColumns, KeyWhitening is applied to the specific bytes before extracting from intermediate rounds. Four round AES ASC-1, operates in a Cipher Feedback (CFB) mode which means that the processing of each plaintext block has to be completed before the processing of the next one starts. Therefore, implementation presented here is sequential. However from Figure 3, parts of implementation could be implemented in parallel architecture.

Systems control unit – The unit is implemented as a finite state machine to supervise the core between AES and ASC-1, generate address for accessing the round keys from the block and handle communication between blocks. The unit generates the signal to notify the external source that a new plaintext may be loaded as soon as core is ready.

Authentication Tag (τ) – Finally the authentication tag is calculated once n numbers of block are encrypted (the maximum number of messages and maximum length to be encrypted is 2^{48}),

4.1 Frame Delay

The end goal of ASC-1 authenticated encryption scheme is to achieve both message secrecy and authenticity in a single cryptographic primitive with the focus to achieve high throughput and minimal overhead for wireless devices. Based on the design of ASC-1, two different approaches are proposed – Key setups during transmission or parallel key setup with the encryption core.

As shown in Figure 4, when the frames passes through the core, only payload have to be encrypted and rest remains in the plaintext. However to initiate the encryption of the payload requires Initial phase generation i.e. calculating initialization vector and keys for the encryption based on Counter $(Cntr)$ and Master Key (K_M). The process is repeated for every frame, where Counter values are varied but Master Key remains same for the session.

In the first approach, the key setup is triggered at the start of the transmission. The unencrypted prefix (header) of the frame is validated and passed though the bypass unit and waits for the encrypted and authenticated pay-

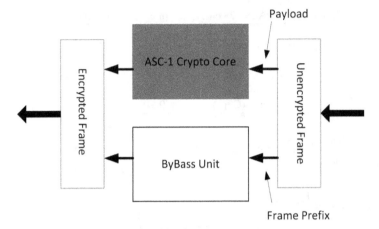

Figure 4 Crypto architecture.

load. Once the encryption is done, the whole frame is packed and sent to the transmitter. During the transmission of the frame, key setup for next frame is performed and stored in the logic. Based on previous sections, two hardware architectures were investigated: basic iterative and parallel architecture. Depending on the area constraints and acceptable delay for the specific applications, either of the architecture for initial phase can be chosen. Based on our results, iterative architecture has a latency of 248 ns, whereas parallel architecture takes about half the time but three times in area. However in either of the architectures this approach may cause some minor end-to-end delays.

To overcome these delays, keys can also be computed in parallel with the encryption core. In this approach, initial phase is generated before the start of the transmission and keys are stored in the internal logic. For subsequent frames new keys are generated in parallel with the encryption core processing last block of the previous frame. This approach may not cause any delays but it comes with the cost of high area consumption.

5 Implementation Results

The results of hardware implementation of "ASC-1: An Authenticated Encryption Stream Cipher" are tabulated in this section. ASC-1 is implemented in VHDL and the target device is Xilinx Spartan-3 sxc3s700an FPGA. The software used for this design is Xilinx ISE-12.4. This is used for writing,

Table 1   Performance of AES-128 encryption in parallel and iterative architecture.

AES-128 Encryption		
Performance	Iterative	Parallel
Number of Slices	1736	15550
Number of Clock Cycles	62	30
Latency (ns)	248	120
Throughput (Gbps)	0.516	up to 32

debugging and optimizing, and all the simulations are carried out in ISim simulator.

5.1 ASC-1 Performance

In an ASC-1 scheme, the underlying block cipher, i.e. AES, is used only in the forward encryption direction for both ASC-1 encryption and decryption. This characteristic make ASC-1 an attractive candidate for hardware where area is limited. Each round in the scheme consists of four basic transformations, i.e., SubBytes, ShiftRows, MixColumns and AddRoundKey. The S-Box byte substitution function can be implemented either by using combinational logic or using a 256 × 8 bit look-up table, using ROM (Read Only Memory). Use of ROM is the most optimal implementation in terms of area/performance – in an FPGA. To access ROM, inputs used as addresses and output is acquired at the data out bus. A state matrix consists of 16 bytes and for each byte substitution 16 ROMs have to be used. FPGA used in this implementation Xilinx Spartan-3AN provides fast on-chip memories, called BlockRAMs. BlockRAMs can be configured as dual port ROMs. This reduces the amount of ROMs in half, i.e. 8. This whole process requires only one clock cycle. Other three transformations during the encryption/decryption process are basic operations and takes minimal resources.

Table 1 presents the detail implementation results for the AES-128 encryption system. AES encryption is used during the initial phase i.e., for the calculation of IV and keys used for encryption and authentication of data. Same key is used to encrypt all the initial values in ECB non-feedback mode. With encryption in non-feedback mode, processing of data blocks can be performed independently from other blocks and all the blocks can be encrypted in parallel. Following table shows the throughput, latency and area used for parallel and iterative hardware architectures. The system is set to 250 MHz with a clock cycle of 4 ns.

Table 2 Performance of ASC-1 encryption core iterative architecture.

Performance	Iterative
Number of Slices	1796
Number of Clock Cycles	41
Latency (ns)	164
Throughput (Gbps)	0.780

A huge trade-off between area and performance of the system can be clearly seen. The number of slices used in a parallel architecture is almost nine times as much as in an iterative architecture. However, on the other side, the throughput of the Iterative architecture is much lower than parallel architecture.

Table 2 provides the results of ASC-1 encryption core; the core consists of 4-Round AES and operates in CFB mode to compute an authentication tag over the encrypted message. In feedback modes it is not possible to encrypt next block of data until encryption of previous block is completed. As a result, data blocks must be encrypted sequentially, with no capability of parallel processing.

As compared to AES iterative architecture, data is processed only four times instead of ten times and initial and final rounds are not included. The order of bit transformations inside each round is also different as compared to AES; AddRoundKey transformation is performed at the start of each round unlike AES.

6 Conclusion

In this paper, we presented a single pass authenticated encryption scheme: ASC-1 for wireless reconfigurable chips with the focus to achieve high throughput and low overhead. The goal of this scheme is to address two main security objectives, i.e., Confidentiality and Authenticity. This is achieved by performing both the encryption and message authentication in a single pass as opposed to the traditional approaches, which requires two passes. Additionally, we have designed and implemented ASC-1 authenticated encryption scheme on FPGAs. The crypto module, i.e., ASC-1 is placed on the re-configurable chip is responsible for the confidentiality and integrity of the data flow passing through it from both the sides. We have also explored any possible frame delay due to the initial key setup with every frame. Based on the available resources, two different approaches are proposed.

After analyzing the performance parameters, we conclude that ASC-1 is suitable for low-cost low-power reconfigurable wireless devices with negligible or no delays. The resulting implementation consumes moderate number of slices on FPGA and achieves throughput in the range of 0.8 Gbps. Comparing with traditional two pass approaches, the presented design demonstrates high throughput and small area to performance ratio.

References

[1] D. Ma and G. Tsudik. Security and privacy in emerging networks. IEEE Wireless Communications, 17(5), 12–21, October 2010.
[2] C. Jutla. Encryption modes with almost free message integrity. In Advances in Cryptology EUROCRYPT 2001, Lecture Notes in Computer Science, Vol. 2045, pp. 529–544. Springer Verlag, Berlin, 2001.
[3] V.D. Gligor and P. Donescu. Fast encryption and authentication: XCBC encryption and XECB authentication modes. In Proceedings of Fast Software Encryption 2001, M. Matsui (Ed.), Lecture Notes in Computer Science, Vol. 2355. Springer Verlag, Berline, 2001.
[4] P. Rogaway, M. Bellare, J. Black, and T. Krovetz. OCB: A block-cipher mode of operation for efficient authenticated encryption. In Proceedings of 8th CCS. ACM, New York, 2001.
[5] M. Bellare and C. Namprempre. Authenticated encryption: Relations among notions and analysis of the generic composition paradigm. In Advances in Cryptology – ASIACRYPT 2000, Vol. 1976, pp. 531–545. Springer Verlag, Berlin, 2000.
[6] G. Jakimoski and S. Khajuria. ASC-1: An authenticated encryption stream cipher. In Selected Areas in Cryptography 2011, Lecture Notes in Computer Science, Vol. 7118, pp. 356–372. Springer Verlag, Berlin, 2011.

Biographies

Samant Khajuria is a PhD student at the Center for Tele Infra Structure (CTIF) Copenhagen at Aalborg University (Denmark). He received his Bachelor in Electronics and Communication in 2006 from PES Institute of Technology – Bangalore (INDIA) and Masters Degree in Communication networks (specializing in security) in 2008 from Aalborg University Copenhagen. He started as a research assistant at the Center for Wireless Systems and Applications (CWSA), before starting his PhD. Major research areas include Cryptography, Cognitive Radio, Computer Networks, FPGAs.

Birger Andersen is a Professor at Copenhagen University College of Engineering, Denmark, and Director of Center for Wireless Systems and Applications (CWSA) related. He received his M.Sc. in Computer Science in 1988 and his Ph.D. in Computer Science in 1992, both from University of Copenhagen. He was an assistant professor at University of Copenhagen, a visiting professor at Universität Kaiserslautern, Germany, and an associate professor at Aalborg University. Later he joined the IT Department of Copenhagen Business School, Denmark, and finally Copenhagen University College of Engineering. He is currently involved in research in wireless systems with a focus on security.

Activity Modelling and Comparative Evaluation of WSN MAC Security Attacks

Pranav M. Pawar[1], Rasmus H. Nielsen[2], Neeli R. Prasad[2],
Shingo Ohmori[3] and Ramjee Prasad[1]

[1] *Center for TeleInFrastruktur, Aalborg University, Aalborg, Denmark;*
e-mail: {pmp, prasad}@es.aau.dk,
[2] *Center for TeleInFrastruktur, Princeton, USA; e-mail: {rhn, np}@es.aau.dk*
[3] *Center for TeleInFrastruktur, Yokosuka, Japan; e-mail: shingo_o@yiai.jp*

Abstract

Applications of wireless sensor networks (WSNs) are growing tremendously in the domains of habitat, tele-health, industry monitoring, vehicular networks, home automation and agriculture. This trend is a strong motivation for malicious users to increase their focus on WSNs and to develop and initiate security attacks that disturb the normal functioning of the network in a severe manner. Such attacks affect the performance of the network by increasing the energy consumption, by reducing throughput and by inducing long delays. Of all existing WSN attacks, medium access control (MAC) layer attacks are considered the most harmful as they directly affect the available resources and thus the nodes' energy consumption.

The first endeavour of this paper is to model the activities of MAC layer security attacks to understand the flow of activities taking place when mounting the attack and when actually executing it. The second aim of the paper is to simulate these attacks on hybrid MAC mechanisms, which shows the performance degradation of a WSN under the considered attacks. The modelling and implementation of the security attacks give an actual view of the network which can be useful in further investigating secure mechanisms to reduce the degradation of the performance in WSNs due to an attack. Lastly, the paper proposes some solutions to reduce the effects of an attack.

Journal of Cyber Security and Mobility, Vol. 1, 205–225.

Keywords: wireless sensor networks (WSNs), media access control (MAC), activity modelling, security attacks.

1 Introduction

A WSN consists of small sensor nodes, each one equipped with limited battery, a microprocessor, a small amount of memory and a transducer. WSNs are versatile networks with a very wide domain of applications and their resource-constrained nature is an important research challenge. Like all other networks, WSN resources are mainly affected by the MAC layer and MAC layer protocols play an important role in resource utilisation, network delays, scalability and energy consumption [1].

Due to the rise of many mission critical WSN applications, another great challenge is security. The range and number of security attacks in WSNs have increased significantly over the last decade [2] and it is therefore necessary to design WSNs and related protocols also considering constraints with respect to security. Attacks can happen at all layers of a WSN but are more harmful when they are in the form of resource consumption attacks. Resource consumption attacks mainly take place at the MAC layer because this is the layer that controls the access to the resources in the network.

This paper focuses on denial of service MAC layer attacks in WSNs. Denial of service attacks in WSNs are primarily affecting the sleep mode of WSN nodes, which is often referred to as denial of sleep. During sleep mode, the nodes save energy by keeping the radio off and denial of sleep attacks prevent nodes from going into this mode, which increases the energy consumption and also reduces the total network lifetime [3, 4].

The objectives of this paper are to analyse the activities taking place in order to carry out the attack, to investigate how the attack can be implemented and to propose a solution that can reduce the effects of the attacks. The attacks are modelled using activity modelling, which is an efficient tool to understand the flow of activities during the implementation of the tasks. The attacks are implemented using the tool Network Simulator 2 (NS-2) considering hybrid MAC mechanisms [5] and the results show the actual performance degradation due to the attacks. The results are useful for proposing secure hybrid WSN MAC mechanisms and the paper proposes a novel solution that reduces the effect of a specific attack.

The remainder of this paper is organised as follows: Section 2 gives an overview of MAC layer denial of service/sleep attacks. Section 3 presents the activity modelling of security attacks. Section 4 discusses the simulation of security attacks and the results obtained and Section 5 explains the proposed solutions to reduce the effect of the attacks. Finally, Section 6 concludes the paper with future work.

2 MAC Layer Security Attacks

2.1 Collision Attack

The malicious collision attack [6, 7] can be easily launched by a compromised sensor node, which does not follow the MAC protocol rules and thereby causes collisions with neighbouring nodes' transmissions by sending short noise packets. This can cause a lot of disruptions to the network operation and will lead to retransmissions and wasted energy. The attack does not consume much energy of the attacker and it is difficult to detect because of the broadcast nature of the wireless environment.

2.2 Unintelligent Replay Attack

In case of the unintelligent replay attack [3], the attacker does not have MAC protocol knowledge and no ability to penetrate the network. Here, recorded events are replayed into the network which prevent nodes from entering sleep mode and lead to waste in energy in receiving and processing the extra packets. If nodes are not equipped with an anti-replay mechanism, this attack causes the replayed traffic to be forwarded through the network, consuming energy at each relaying node on the path to the destination. The replaying of events has adverse effect on the network lifetime and overall performance of a WSN.

2.3 Unauthenticated Broadcast Attack

In an unauthenticated broadcast attack [3], the attacker has full knowledge of the MAC protocol but does not have the capability to penetrate the network. Here, the attacker broadcasts unauthenticated traffic into the network by following all MAC rules and this disrupts the normal sleep and listen cycle of the node and places most of the nodes in listen mode for an extended amount of time, which leads to increased energy consumption and reduction in network lifetime. Such attacks cause severe harm on MAC protocols by producing short control or data messages and have short adaptive timeout period.

2.4 Full Domination Attack

Here, the attacker has full knowledge of the MAC layer protocol and has the ability to penetrate the network. This type of attack is one of the most destructive to a WSN as the attacker has the ability to produce trusted traffic to gain the maximum possible impact from denial of sleep. This attack is carried out using one or more compromised nodes in the network and all MAC layer protocols are vulnerable to this kind of attack [3].

2.5 Exhaustion Attack

The attacker who commences an exhaustion attack [3] has knowledge about the MAC protocol and the ability to penetrate the network. These attacks are possible only in case of request to send (RTS)/clear to send (CTS) based MAC protocols. In this attack, the malicious node sends RTS to a node and if the node replies with CTS, the malicious node will repeatedly transmit the RTS to the node, which will prevent the node from going into sleep mode and instead drain the energy of the node. This attack is affecting the node lifetime and can partition the network, which lead to reduced network lifetime.

2.6 Intelligent Jamming Attack

The intelligent jamming attack is one of the most severe attacks and in this attack the attacker has full protocol knowledge but does not have the ability to penetrate the network. The attacker injects unauthenticated unicast and broadcast packets into the network. These attacks can differentiate between control and data traffic and unlike the unauthenticated replay attack it selectively replays the data or control packets [3, 8].

3 Activity Modelling of MAC Security Attacks

3.1 Activity Modelling

Activity diagrams are often used to give a functional view of a system as it describes logical processes, or functions, where each process describes a sequence of tasks and the decisions that govern when and how they are performed. UML [9–11] is a tool that provides functional modelling in the form of an activity diagram, which is designed to support the description of behaviours that depend upon the results of internal processes, as opposed to external events as in interaction diagrams. The flow in an activity diagram is driven by the completion of an action. Activity diagrams are useful tools

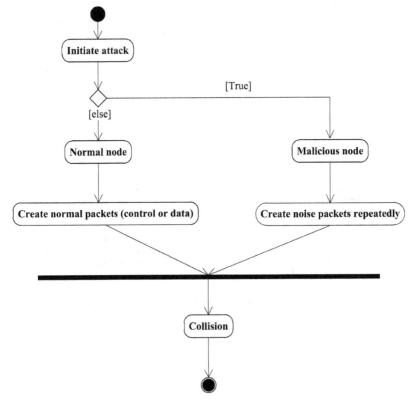

Figure 1 Activity modelling of collision attack.

to understand the basic flow of security attacks and will be utilised in the following to do so.

3.2 Activity Modelling of Security Attacks

For all of the presented attacks, the external attacker will have to initiate an attack by utilising an exploit on a vulnerable normal node to turn it into a malicious node. If the attacker succeeds in initiating the attack the normal node becomes malicious and otherwise it continues to operate as a normal node.

3.2.1 Collision Attack

Figure 1 shows the activity diagram for the collision attack and the different activities are as follows:

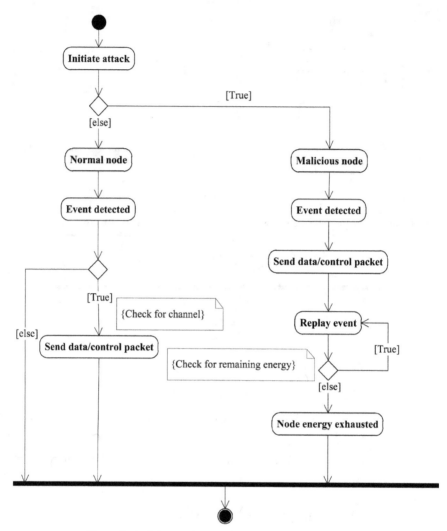

Figure 2   Activity modelling of unintelligent replay attack.

- The malicious node randomly creates noise packets and transmits them into the network.
- A normal node starts a transmission to the sink either by direct communication or through relays using multi-hop communication.

- A collision happens between the control or data packet from the normal node and the noise packet from the malicious node. Repeatedly collisions will reduce the performance of the network.

3.2.2 Unintelligent Replay Attack

The sequence of activities in case of an unintelligent replay attack is shown in Figure 2 and are as follows:

- The normal node has data to send and checks if the channel is available and, if it is, the node starts the transmission.
- The malicious node records the transmission as if in normal node mode, which it keeps replaying unintelligently, i.e. without making differentiation between data and control packets; it will replay any transmission the normal node would have generated.
- The malicious node checks the remaining energy on each replay and once the energy is exhausted, the attack will be terminated and the external attacker will try to initiate the attack on another node.

3.2.3 Unauthenticated Broadcast Attack

Figure 3 shows the activity modelling of the unauthenticated broadcast attack. The sequence of activities performed by the normal and malicious node is as follows:

- The normal node does communication as in the previous attack.
- The malicious node uses similar transmissions, but broadcasts the packet to all nodes in the network and, further, tries to authenticate itself, which fails.
- If the broadcast takes place during transmission of a normal node a collision will take place. These collisions and the failed attempt to authenticate lead to performance degradation and thereby excessive energy consumption.

3.2.4 Full Domination Attack

The modelling of the sequence of activities for the full domination attack can be seen from Figure 4 and activities are as follows:

- The normal node broadcasts a packet into network, if the channel is available.
- The malicious node does the same and tries for authentication. As the attacker has full network knowledge, the authentication is successful

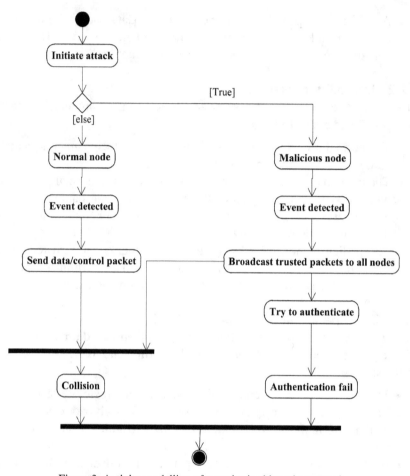

Figure 3   Activity modelling of unauthorised broadcast attack.

and the malicious packet is transmitted while the malicious node tries to introduce collisions during normal traffic.

• The malicious node can also replay the communications unintelligently and broadcast it until the node's energy is exhausted. The full domination attack reduces the efficiency of the network by introducing authenticated broadcast and by replaying transmissions.

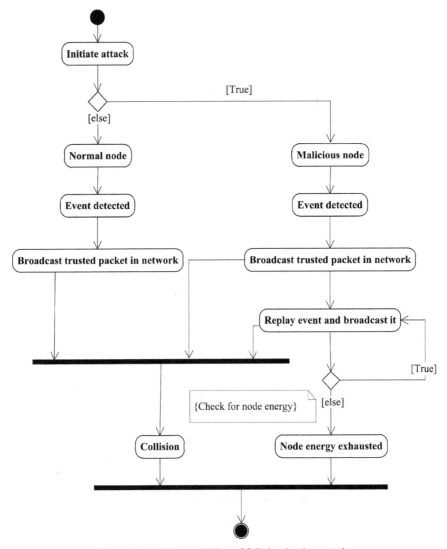

Figure 4   Activity modelling of full domination attack.

3.2.5 Exhaustion Attack

Figure 5 explains the sequence of activities during an exhaustion attack and the sequence of activities can be explained as follows:

- The normal node can send RTS, receive CTS from destination and send data towards the destination node.

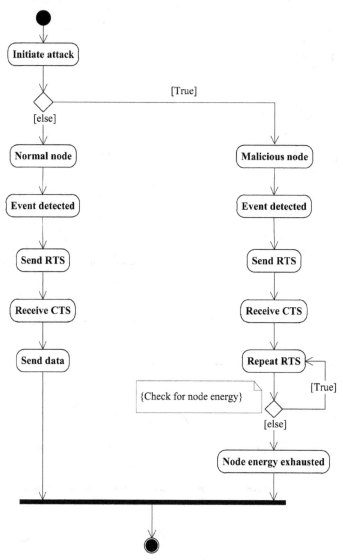

Figure 5 Activity modelling of exhaustion attack.

- In case of the malicious node, it sends RTS and waits for CTS from the destination node. If it receives the CTS, it will send the RTS repeatedly towards the destination node until its energy is exhausted.

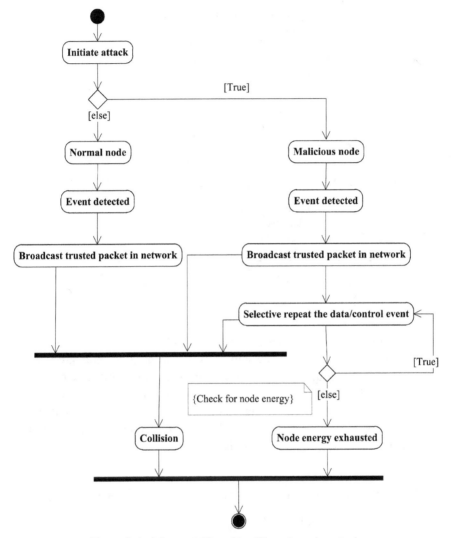

Figure 6 Activity modelling of intelligent jamming attack.

3.2.6 Intelligent Jamming Attack

Figure 6 shows the sequence of activities that happen during an intelligent jamming attack and the activities are as follows:

- The normal node has data to send and broadcasts it if the channel is available.

Table 1 Simulation and node parameters.

Parameter Name	Setting Used
Wireless Physical	
Network Interface Type	Wireless Physical
Radio Propagation Model	Two-Ray Ground
Antenna Type	Omni-directional antenna
Channel Type	Wireless Channel
Link Layer	
Interface Queue	Priority Queue
Buffer Size of IFq	50
MAC	Z-MAC
Routing Protocol	Ad hoc Routing
Energy Model	
Initial Energy (initialEnergy_)	100 J
Idle Power (idlePower_)	14.4 mW
Receiving Power (rxPower_)	14.4 mW
Transmission Power (txPower_)	36.0 mW
Node Placement	
Number of Nodes	100
Node Placement	Random

- The malicious node does the same, authenticates to the node and broadcasts the packet in the same way as for the full domination attack.
- The most important feature of the intelligent jamming attack is its intelligent behaviour. It can differentiate between data and control packets, and will selectively replay the events until the node energy is exhausted.
- The replaying of event and broadcast of authenticated packets lead to collisions during normal transmissions.

4 Simulation of Security Attacks on Hybrid WSN MAC

4.1 Simulation Details

All simulations are carried out using the discrete event simulator NS-2 and the simulation parameters are shown in Table 1. The idle power, receiving power, transmission power and sleep power are considered according to the RFM TR 3000 transceiver. The simulations are performed using the hybrid MAC mechanism Zebra-MAC (ZMAC) [11] and the simulated scenarios are:

- ZMAC without any attacks.
- ZMAC under unintelligent replay attack.
- ZMAC under unintelligent broadcast attack.

- ZMAC under exhaustion attack.
- ZMAC under collision attack.
- ZMAC under full domination attack.
- ZMAC under intelligent jamming attack.

The simulations are carried out under the assumption that the attacker can initiate the attack from multiple nodes. The initial simulation is done using four malicious nodes, but the impact of varying malicious nodes (from 2 to 32) is also investigated.

4.2 Results and Discussion

Figures 7–9 show the performance, i.e. energy consumption, throughput and delay of ZMAC under normal conditions and under attacks. The figures show that the performance of the WSN degrades with the attacks and the reasons for the performance degradations under the individual attacks are as follows:

- Unauthenticated Broadcast Attack: The performance degradation due to this attack is less compared to the other attacks because this attack utilises more energy and requires extra time for authentication of the broadcast packets coming from the malicious nodes. The attack results in degradation of the total throughput of the network due to an increased number of collisions caused by the unauthenticated packets and the following retransmission of packets from trusted nodes.
- Unintelligent Replay Attack: The performance degradation due to this attack is more severe than for the unauthenticated broadcast attack because this attack increases the energy consumption by replaying data or control packets and thereby wastes energy. A significant increase in delay can be observed because the node checks for energy at each replay and also requires additional time to carry out the replay. This unnecessary replay keeps the channel busy, which may introduce collisions and prevent transmission of other packets, which result in degradation of the total throughput of the network. The attack has more severe performance degradation than the unintelligent replay attack because it can take place in any situations, i.e. (i) no protocol knowledge, no ability to penetrate, (ii) full protocol knowledge, no ability to penetrate, and (iii) full protocol knowledge, network penetrated.
- Exhaustion Attack: The most adverse effect of this attack is that it totally blocks the transmission towards one particular node and blocks this node until its energy is depleted or the network becomes partitioned.

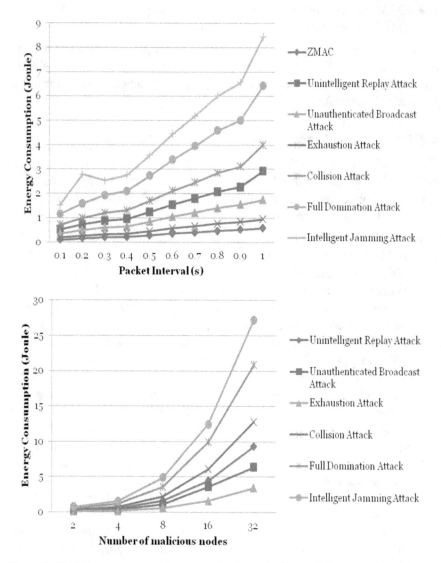

Figure 7 (Top) Energy consumption vs. packet interval. (Bottom) Energy consumption vs. number of malicious nodes.

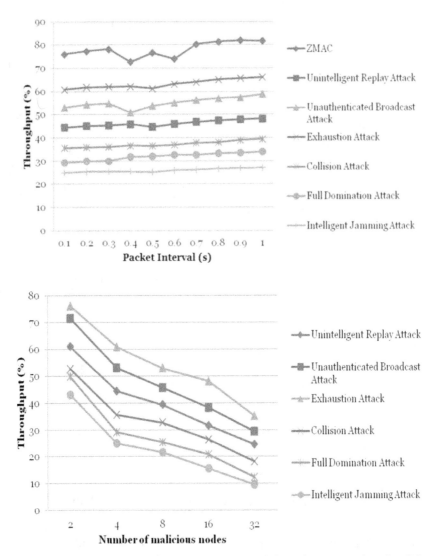

Figure 8 (Top) Throughput vs. packet interval. (Bottom) Throughput vs. number of malicious nodes.

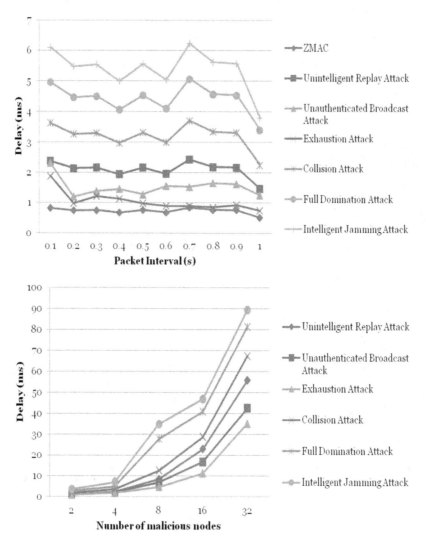

Figure 9 (Top) Delay vs. packet interval. (Bottom) Delay vs. number of malicious nodes.

- Collision Attack: The noise packets introduced by this attack result in significant performance degradation due to the increased number of collisions in the network. The collisions result in performance degradation by (i) blocking the channel, (ii) increasing the retransmission of packets, (iii) introducing delays, and (iv) reducing the chances of packets to reach their destination. The effect of collisions is adverse as the traffic load increases.

- Full Domination Attack: This attack is a combination of the previously two attacks and therefore has more adverse effects. The results show that energy, throughput and delay degradations are much increased compared to the previous attacks as this attack increases the delay and energy consumption by repeatedly broadcasting packets. This makes the channel constantly busy, so it will not be available other nodes to transmit and it also reduces the throughput by not giving the chance to new packets to be transmitted through the network. As for the exhaustion attack, it also partition the network but much faster than the previous attacks.

- Intelligent Jamming Attack: This is the most disastrous of the considered attacks because it works intelligently by selectively retransmitting data and control packets. It requires in-depth knowledge of the protocols used in the network. The results show that the performance degradation of this attack is slightly more severe than the full domination attack as this attack cannot intelligently retransmit.

5 Proposed Solutions

The implementation shows the adverse effects on the network of the listed WSN MAC layer attacks. The attacks degrade the performance of the WSN with 50% or more and to reduce these effects some possible solutions are:

- Cluster-based MAC protocols: Cluster-based MAC protocols improve the scalability of the network by stabilising the network topology at the level of sensors and thus lower the topology maintenance overhead. The clustering can also reduce the number of required slots by increasing the reuse of slots that, in turn, can reduce the amount of delay in communication. One important advantage offered by cluster-based protocols is their inherent security, as an attack on a cluster-based MAC protocols will be bound to that particular cluster and not affect the whole network. This automatically reduces the overall impact of the attack on the total performance of the WSN.

- Secure slot assignment: Secure slot assignment assumes that some slots are secure and are given only to nodes that are transmitting sensitive information. Nodes with secure slots assigned can start communication with another node by checking the identity of the node to determine if it is secure to communicate with or not. This will help to reduce the influence of MAC security attacks in terms of reducing energy consumption, delay and increasing throughput.

6 Conclusion

The activity modelling of WSNs MAC layer security attacks gives a detailed view of activities executed during mounting of the attacks, which is essential knowledge for proposing more efficient security mechanisms that can withstand the attacks. The paper also provides simulation results of security attacks on a hybrid MAC mechanism and the results show the trend of network degradation due to the security attacks under varying traffic and number of malicious nodes. Intelligent jamming attacks pose the greatest threat to a WSN because of its intelligent nature, i.e. the attacker has full knowledge of the protocols used, it can easily differentiate between control and data packets and it can penetrate the network. The simulation results in general give a strong motivation and modelling framework for further investigating efficient and secure MAC protocols for WSN.

References

[1] Bachir, A., Dohler, M., Watteyne, T., and Leung, K.K., MAC essentials for wireless sensor networks. IEEE Communications Surveys & Tutorials, 12(2):222–248, 2010.

[2] Chen, X., Makki, K., Yen, K., and Pissinou, N. Sensor network security: A survey. IEEE Communications Surveys & Tutorials, 11(2):52–73, 2009.

[3] Raymond, D.R., Marchany, R.C., Brownfield, M.I., and Midkiff, S.F. Effects of denial-of-sleep attacks on wireless sensor network MAC protocols. IEEE Transaction on Vehicular Technology, 58(1):367–380, January 2009.

[4] Brownfield, M., Gupta, Y., and Davis IV, N. Wireless sensor network denial of sleep attack. In Proceedings of IEEE Workshop on Information Assurance and Security, United States Military Academy, West Point, NY, 15-17 June, pp. 356–364, 2005.

[5] Pawar, P.M., Nielsen, R.H., Prasad, N.R., Ohmori, S., and Prasad, R. Hybrid mechanisms: Towards an efficient wireless sensor network medium access control. In Proceedings of WPMC, Brest, France, 3–6 October, pp. 492–496, 2011.

[6] Reindl, P., Nygard, K., and Xiaojiang, Du. Defending malicious collision attacks in wireless sensor networks. In Proceedings EUC, Hong Kong, China, 11–13 December, pp. 771–776, 2010.

[7] Ren, Q. and Liang, Q. Secure Media Access Control (MAC) in wireless sensor network: Intrusion detections and countermeasures. In Proceedings PIMRC, Berlin, Germany, 5–8 September, pp. 3025–3029, 2004.

[8] Law, Y.W., Palaniswami, M., Van Hoesel, L., Doumen, J., Hartel, P., and Havinga, P. Energy-efficient link-layer jamming attacks against wireless sensor network MAC protocols. ACM Transactions on Sensor Networks, 5(1):6.1–6.38, February 2009.

[9]   Peder, T. UML Bible, First Edition. John Wiley & Sons, 2003.

[10] Pawar, P.M., Nielsen, R.H., Prasad, N.R., Ohmori, S., and Prasad, R. Behavioural modelling of WSN MAC layer security attacks: A sequential UML approach. Journal of Cyber Security and Mobility, 1(1):65–82, January 2012.

[11] Hong, S. and Lim, S. Analysis of attack models via unified modelling language in wireless sensor networks: A survey study. In Proceedings WCNIS, Beijing, China, 25–27 January, pp. 692–696, 2010.

[12] Rhee, I., Warrier, A., Aia, M., Min, J., and Sichitiu, M.L. ZMAC: A hybrid MAC for wireless sensor networks. IEEE/ACM Transactions on Networking, 16(3):511–524, June 2008.

Biographies

Pranav M. Pawar graduated in Computer Engineering from Dr. Babasaheb Ambedkar Technological University, Maharashtra, India, in 2005 and received a Master in Computer Engineering from Pune Universiry, in 2007. From 2006 to 2007, was working as System Executive in POS-IPC, Pune, India. From January 2008, he has been working as an Assistant Professor in Department of Information Technology, STES's Smt. KashibaiNavale College of Engineering, Pune. Currently he is working towards his Ph.D. in Wireless Communication from Aalborg University, Denmark. He published 17 papers at national and international level. He is IBM DB2 and IBM RAD certified professional. His research interests are Energy efficient MAC for WSN, QoS in WSN, wireless security, green technology, computer architecture, database management system and bioinformatics.

Rasmus Hjorth Nielsen is an Assistant Professor at Center for TeleIn-Frastruktur (CTIF) at Aalborg University (AAU), Denmark and is currently working as a senior researcher at CTIF-USA, Princeton, USA. He received his M.Sc. and Ph.D. in electrical engineering from Aalborg University in 2005 and 2009 respectively. He has been working on a number of EU and industrial funded projects primarily within the field of next generation networks where his focus is currently security and performance optimization. He has a strong background in operational research and optimization in general and has applied this as a consultant within planning of large-scale networks. His research interests include IoT, WSNs, virtualization and other topics related to next generation converged wired and wireless networks.

Neeli Rashmi Prasad, Ph.D., IEEE Senior Member, Head of Research and Coordinator of Themantic area Network without Borders, Center for TeleInfrastruktur (CTIF), Aalborg University, Aalborg, Denmark. Director of CTIF-USA, Princeton, USA and leading IoT Testbed at Easy Life Lab and Secure Cognitive radio network testbed at S-Cogito Lab. She received her Ph.D. from University of Rome "Tor Vergata", Rome, Italy, in the field of "adaptive security for wireless heterogeneous networks" in 2004 and an M.Sc. (Ir.) degree in Electrical Engineering from Delft University of Technology, the Netherlands, in the field of "Indoor Wireless Communications using Slotted ISMA Protocols" in 1997. During her industrial and academic career for over 14 years, she has lead and coordinated several projects. At present, she is leading a industry-funded projects on Security and Monitoring (STRONG) and on reliable self organizing networks REASON, Project Coordinator of European Commission (EC) CIP-PSP LIFE 2.0 for 65+ and social interaction and Integrated Project (IP) ASPIRE on RFID and Middleware and EC Network of Excellence CRUISE on Wireless Sensor Networks. She is co-caretaker of real world internet (RWI) at Future Internet. She has lead EC Cluster for Mesh and Sensor Networks and Counsellor of IEEE Student Branch, Aalborg. She is Aalborg University project leader for EC funded IST IP e-SENSE on Wireless Sensor Networks and NI2S3 on Homeland and Airport security and ISISEMD on telehealth care. She is also part of the EC SMART Cities workgroup portfolio. She joined Libertel (now Vodafone NL), Maastricht, the Netherlands as a Radio Engineer in 1997. From November 1998 till May 2001, she worked as Systems Architect for Wireless LANs in Wireless Communications and Networking Division of Lucent Technologies, Nieuwegein, the Netherlands. From June 2001 to July 2003, she was with T-Mobile Netherlands, The Hague, the Netherlands as Senior Architect for Core Network Group. Subsequently, from July 2003 to April 2004, she was Senior Research Manager at PCOM:I3, Aalborg.

Shingo Ohmori is the president of YRP International Alliance Institute and professor and president of CTIF-Japan, Aalborg University, Denmark. He received the B.E., M.E., and Ph.D. degrees in electrical engineering from the University of Tohoku, Japan, in 1973, 1975, and 1978, respectively. In 1978, he joined National Institute of Information and Communications Technology (NICT), and he resigned a Vice President of NICT in March 2009. During 1983–1984, he was a visiting research associate at the ElectroScience Laboratory, the Ohio State University, Columbus, Ohio. He is author of *Mobile Satellite Communications* (Artech House, 1998), a co-author of *Mobile Antenna Systems Handbook* (Artech House, 1994) and Co-Editor of

Towards Green ICT (River Publishers, 2010). He was awarded the Excellent Research Prize from the Minister of Science and Technology Agency of Japan in 1985, and the Excellent Research Achievements Prize of the IEICE in 1993. He was a guest professor of Yokohama National University, and now a guest professor of Dalian Technical University and Beijing University of Posts and Telecommunications, China. Dr. Ohmori is an IEEE Fellow and an IEICE Fellow (Institute of Electronics, Information and Communication Engineers, Japan).

Ramjee Prasad (R) is currently the Director of the Center for TeleInfrastruktur (CTIF) at Aalborg University, Denmark and Professor, Wireless Information Multimedia Communication Chair.

He is the Founding Chairman of the Global ICT Standardisation Forum for India (GISFI: www.gisfi.org) established in 2009. GISFI has the purpose of increasing of the collaboration between European, Indian, Japanese, North-American, and other worldwide standardization activities in the area of Information and Communication Technology (ICT) and related application areas. He was the Founding Chairman of the HERMES Partnership – a network of leading independent European research centres established in 1997, of which he is now the Honorary Chair.

Ramjee Prasad is the founding editor-in-chief of the Springer *International Journal on Wireless Personal Communications*. He is member of the editorial board of other renowned international journals, including those of River Publishers. He is a member of the Steering, Advisory, and Technical Program committees of many renowned annual international conferences including Wireless Personal Multimedia Communications Symposium (WPMC) and Wireless VITAE. He is a Fellow of the Institute of Electrical and Electronic Engineers (IEEE), USA, the Institution of Electronics and Telecommunications Engineers (IETE), India, the Institution of Engineering and Technology (IET), UK, and a member of the Netherlands Electronics and Radio Society (NERG) and the Danish Engineering Society (IDA). He is also a Knight ("Ridder") of the Order of Dannebrog (2010), a distinguishment awarded by the Queen of Denmark.

Cooperative Wireless Communications and Physical Layer Security: State-of-the-Art

Vandana Milind Rohokale, Neeli Rashmi Prasad and Ramjee Prasad

Center for TeleInFrastruktur, Aalborg University, Aalborg, Denmark;
e-mail: {vmr, np, prasad}@es.aau.dk

Abstract

One morning, we were waiting for our college bus. The Wipro industry bus was slowly passing nearby us looking for its employees. At the last moment, when the driver increased speed, one person stepped down from the auto rickshaw and shouted "stop the bus, stop the bus". Voluntarily, whoever was present started shouting "stop, stop the bus". The sound finally reached the bus driver who stopped the bus, and the employee could catch it in time. This analog from everyday realistic life simply depicts the spirit of cooperative wireless communication which utilizes the information overheard by neighbouring nodes to offer reliable communication between sender and receiver. Future converged wireless networks are expected to provide high data rate services with extension in coverage area. Also, the next generation networks should possess bandwidth efficiency, less power consumption ability with small sized mobile equipment. Multiple-input multiple-output (MIMO) system is the best technique for the provision of communication diversity wherein multiple antennas are installed at the sender and receiver. In today's miniaturizing electronics era, the hardware implementation of MIMO in the mobile equipment is not feasible due to resource constraints. Cooperative wireless communication (CWC) is the upcoming virtual MIMO technique to combat fading and achieve diversity through user cooperation. Physical layer security (PLS) is the imminent security guarantee for the cooperative communication.

Journal of Cyber Security and Mobility, Vol. 1, 227–249.

Keywords: Multiple Input Multiple Output (MIMO), Cooperative Wireless Communication (CWC), Physical Layer Security (PLS).

1 Introduction

Currently, wireless communication and mobile computing are the buzz words for the telecom industry. For multimedia applications, the user needs higher data rates at which data transactions can take place efficiently. Gigabit wireless communication is the dream which is being chased by scientists and researchers. Capacity provided by single antenna systems is bounded by the Shannon limit. Diversity gains like capacity and high data rates are possible with the MIMO systems [1]. Diversity is nothing but a mechanism for reliability improvement in the transmitted message signal which makes use of two or more communication channels of different characteristics.

However today's booming wireless techniques, such as adhoc networks, wireless sensor networks and cognitive radio networks, make use of resource constrained miniature devices. The hardware implementation of the MIMO system poses problem due to size, weight and cost [2]. MIMO is the only key solution to bring spectrally efficient Gigabyte wireless communication in reality. Cooperative communication creates the scenario of virtual MIMO by utilizing the group communicating nodes antennas. Cooperation among the network node entities ensures the regulation of the network traffic whereas the traditional multi-hop networks generate contention in the traffic as depicted in Figure 1.

The cooperative broadcasting is prone to eavesdropping attacks due to its multi-node wireless connectivity. Nowadays everybody wishes to use their wireless equipment to make wireless security sensitive transactions like on-line banking, stock trading and shopping. In such cases, the protection of personal and business data is very much important. When a receiver receives a message, it may be concerned about who is the real sender and whether the content of the message has been changed illegally by somebody in the transmission. Message secrecy problem become the important aspect of information security in modern times.

Security of private key cryptosystems depends on secrecy of the secret key. In case of public key systems, it is infeasible to extract private key from the public key. Breaking of a public key is a complex and timely task. Much work is in progress in the direction of enhancement of energy efficiency. But certain issues such as Trusted, Authenticated and Reliable connectivity in multi-node cooperative communication networks in consultation with energy

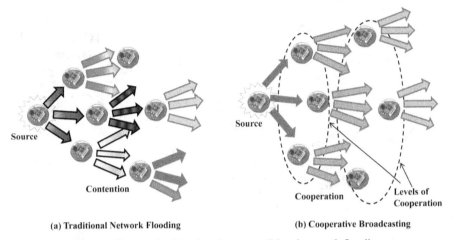

(a) Traditional Network Flooding (b) Cooperative Broadcasting

Figure 1 Cooperative broadcasting vs. traditional network flooding.

efficiency are the real forthcoming challenges. The energy savings in CWC are the result of cross-layer interactive cooperative communication. Routing functions are partially executed in the physical layer.

The traditional cryptographic algorithms namely, AES, DES, and NTRUE etc. include complex mathematical calculations. Since next generation networks like CRNs and WSNs are making use of resource constrained miniature network nodes, these traditional higher layered cryptographic solutions are not feasible for them. Physical layer security employing information theoretic source and channel coding techniques has potential to provide energy efficient security solutions for these networks.

This paper puts forward a physical layer security mechanism for the cooperative networks. In Section 2, related work in the fields of cooperative communication and physical layer security are described. Section 3 depicts the proposed secure cooperative scheme and the different techniques therein. Finally, Section 4 concludes the work.

2 Cooperative Wireless Communication (CWC)

The origin of cooperative communication concept dates back to the research work of Van der Meulen [3] and Cover and Gamal [4], where the concept of relay channel was introduced in between the traditional source-receiver communication. Three node network was considered for the analysis which consists of source, relay and destination. Total system under consideration

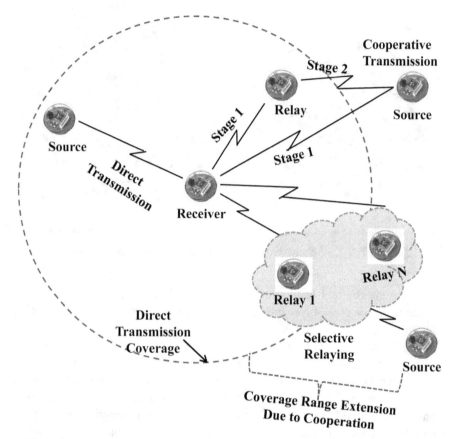

Figure 2 Illustration of direct transmission, cooperative relaying and selective relaying.

has a single bandwidth with broadcasting at the source node and multiple access at the receiver. Cooperation of mobile users with spatial diversity was studied in [5] wherein the network entities transmit their cooperative partner's data with their own data during different time slots. The network node functions as a source for the transmission of their own data and act as a relay for transmission of other nodes' data.

Misha Dohler et al. [6] have appropriately explained cooperation mechanism, wireless relay channel and their modelling, transparent relaying techniques, regenerative relaying techniques and hardware issues in the design of cooperative transceivers for different application scenarios like 3G UMTS Voice/HSDPA Relay and LTE/WiMAX Relay systems. They have also demonstrated some of the real implementations of cooperative diversity mechanisms. Some of the basic advantages of the cooperative relay techniques are depicted in Figure 2 like channel capacity and coverage range improvements.

Cooperative diversity in terms of distributed antenna system was first analyzed in the research work of Saleh et al. [7]. Here two or more information sources form a cooperative group and transmit common information to a single sink. The distributed antenna system was evolved initially for the cellular communication system. With spatial diversity, the advantages of the distributed antenna system are signal strength and channel capacity improvement. For making the transition from a traditional cellular system to a cooperative cellular network, Tao et al. [8] put forth new techniques such as distributed antenna system, multi-cell coordination, group cell mechanism including multiple point transmission and reception (CoMP). These are the stepping stones towards bringing 3GPP LTE-Advanced (LTE-A) into reality.

Research work in [9] has shown that with the coverage range enhancement and improvement in the channel capacity, the cooperative relaying can expressively increase the spectrum efficiency and overall performance of the system. Two intra-cell coordinated multipoint schemes for LTE-Advanced are taken into consideration and it is shown that the network capacity can be considerably improved with the cooperation. Since the transmissions in the cooperative communication are from the nodes at different locations, they may not be time or frequency synchronized. And it becomes difficult to achieve full diversity for the collocated MIMO systems. The work in [10] revisits the techniques for combating the time and frequency asynchronism in one-way as well as two-way cooperative communicating networks.

Reliable communication with reduced energy consumption is the hot issue in the resource constrained networks like WSN and CRN. Cross layer cooperation is the best suited solution for achievement of energy efficiency and reliability in wireless communication. In [11], distributed cross layer technique is proposed which makes use of opportunistic relaying mechanism to achieve quality of service (QoS) in the cooperative communications. Energy savings and low bit error rates can be achieved with the help of cross layer cooperation as shown in Figure 3. Some parts of the routing functions are executed at the physical layer. The diversity provided by the MIMO space time codes can help in the performance improvement at the MAC and upper layers [12]. Liu et al. [13] proposed a new CoopMAC protocol for IEEE 802.11 networks which has inbuilt receiver combining capability. Due to this, the physical layer system cooperates with higher layers in the protocol stack and gets benefits in terms of improvements in the system

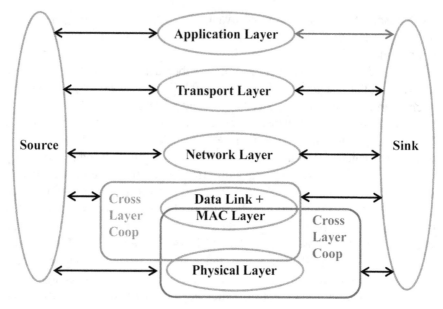

Figure 3 Cross layered cooperative communication [12].

robustness, throughput, delay and interference reduction with the coverage range extension.

The research work of Chen et al. [14] puts forth a new distributed weighted cooperative routing algorithm in which relay selection is based on the weights of the relays. The metrics used for the decision of relay weights are residual energy and channel state information (CSI) at each source-relay-receiver link. Here, the authors have made use of Destination Sequenced Distance Vector (DSDV) routing protocol with consideration of the difficulties due to time synchronization and data packet reduplication. To achieve energy efficient long range communication, cooperative beamforming is the ultimate solution. In the work of Dong et al. [15], a cross layer framework for cooperative communication is proposed which brings the concept of cooperative beamformimg. Here, the cooperative beamforming mechanism is applied for the analysis of the spectrum efficiency of the cooperative communication system. For the study of delay characteristics of the source messages, queuing theory is used.

Independent paths in between source and sink are generated by introducing relay channel in between them in the cooperative communication paradigm. Based on how the signal received form the source is processed at

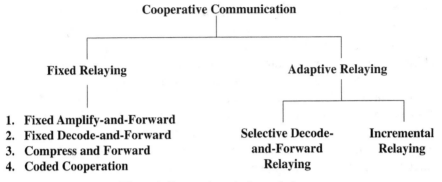

Figure 4 Cooperative relaying techniques.

relay, there are different cooperative communication protocols. Main classes include fixed and adaptive relaying mechanisms as depicted in Figure 4. In case of fixed relaying, the channel resources are distributed in between source and relay in deterministic way. All four techniques under the fixed relaying category work in the predefined deterministic or fixed manner [16]. Adaptive relaying technique containing selective and incremental relaying mechanisms has inbuilt flexibility in the sense that during the adverse conditions like severe channel fading or low SNR conditions, the relay can idle itself.

For optimum diversity gains, proper relay selection plays a vital role in the cooperative communication. The research work of Abdulhadi et al. [17] presents a survey of the distributed relay selection schemes for adhoc cooperative wireless networks. These relay selection schemes include opportunistic relaying, power aware relay selection, switched and examine node selection, opportunistic relaying with limited feedback, simple relay selection, geographical information based relay selection, threshold based relay selection for detect-and-forward, opportunistic AF relaying with feedback, incremental transmission relay selection, outage optimal relay selection, energy efficient relay selection, random priority based relay selection, receive SNR priority list based relay selection, fixed priority transmission protocol, generalized selection combining multiple relay selection scheme and output threshold multiple relay selection. Different performance metrics like objectives, mechanisms, performance, advantages and drawbacks of each of them are illustrated in the tabular way.

Ever-increasing traffic in the vehicular communication systems with safety and spectrum efficiency are the major current issues related to cooperative vehicular communication systems. In [18], the authors have analyzed congestion control and awareness control issues for the cooperative vehicular networks. These critical issues are heavily dependent on the frequency band allocation, medium access control technique being adopted and the availability of the wireless communication technology for particular vehicular communication system. These issues are needed to be researched again.

For the participation in cooperative communication, the network node entities does not get any incentives for their cooperation. Game theoretic cooperation approaches promise to provide proper incentives for the nodes cooperating to relay the information from sender to receiver [19]. Mainly, three types of behaviours are observed in the wireless networks like

1. *No Help (Egoistic Behaviour)*: If the network node has its own data to transmit, then it sends that data independently without any other node's help. If this node does not have its own data for transmission, it remains idle instead of helping other nodes in their data transmissions.
2. *Unidirectional Help (Supportive Behaviour)*: The network node acts only as a supporting relay for the source by helping it to transmit its data towards receiver. No any potential gain is there for such kind of supportive behaviour and that's why it is called as unidirectional help.
3. Mutual Help (Cooperative Behaviour): The network nodes mutually help each other in transmitting their own data as well as their neighbour's data to the intended recipients. The node acts both as source and relay [6].

There should be incentives for the cooperative behaviour and some punishments in terms of costs for the selfish behaviour or no help conditions. Game theory provides the modelling, analysis and solution for the cooperative and non-cooperative behaviours. Three primary techniques are designed for making the provision of cooperation incentives viz. reputation based, resource-exchange-based and pricing based techniques [19].

3 Physical Layer Security (PLS)

The importance of protecting the secrecy of sensitive messages has been realized by people since ancient times. By making use of strong techniques, the storage and transmission of information become cheap and simple in modern times. A huge amount of information is transformed in a way that almost anyone may access it. A lot of new problems related to cryptology appear. For example, an adversary might not only have the means to read transmitted messages, but could actually change them, or the enemy could produce and

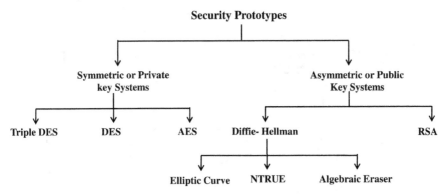

Figure 5 Classification of security mechanisms.

send a false message to the receiver and hope that this would initiate some action. The transactions with the help of wireless networks such as credit card transactions or banking related data exchange communications are prone to the malicious behaviour due to the open nature of wireless medium. Adversaries can easily get access to the wireless transactions and can modify the data therein [20].

Traditional cryptographic techniques include symmetric (private) and asymmetric (public) key systems which are further classified into different mechanisms such as DES, AES, RSA, Diffie–Hellman, etc., as shown in Figure 5. The main problem associated with these cryptographic techniques is that they include complex mathematical calculations which consume considerable part of the resources which are very much crucial for wireless sensor nodes or the consumer radio nodes of the cognitive radio networks.

For establishment of a communication link in between sender and receiver, traditional cryptographic encryption block making use of public and private keys is required at the transmitting end while at the receiving end, channel decoding and decryption blocks are separately used. With the help of information theoretic security, encryption and channel coding blocks are combined in a single secure encoding block as depicted in Figure 6. Also, at the receiving side, channel decoding and decryption blocks are combined to form decoding block. This greatly reduces resource consumption because of relief from the key management functionality which is most costly affair. With the basic wiretap channel, and variants of it like Gaussian, MIMP, compound wiretap, feedback wiretap and wiretap channel with side information are considered for the detail analysis in their work. They have further extended their work from basic wiretap channel to broadcast channels, multiple

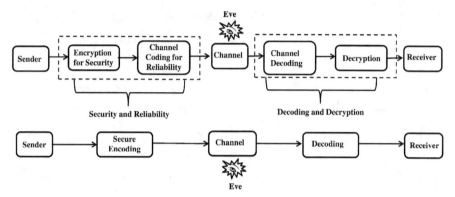

Figure 6 Information theoretic security combines security and reliability functions in a single block.

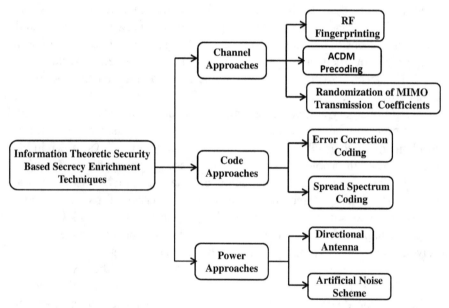

Figure 7 Security enhancement techniques based on information theoretic mechanisms.

access channels, interference channels, relay channels and two-way channels [21].

In the research work of [22], a tutorial is presented on the security improvement techniques at the physical layer in wireless networks. Depending on their characteristic features, these are classified into further subclasses as shown in Figure 7. Two metrics considered for the security analysis include secret channel capacities and computational complexities in comprehensive key search.

Built in physical layer security is nothing but the capacity of the transmission channel. It is called secret channel capacity. It is defined as information theoretic secrecy because the eavesdropper's received signal does not give any information about the original transmitted signal. It just keeps on purely guessing. Information theoretic secrecy is in fact equivalent to perfect secrecy. In the research work of Rohokale et al. [23], cooperative jamming technique is used to confound the eavesdropper and relay is cooperatively sending jamming signals towards the eavesdropper to confuse about the actual transmitted signal. Here, the secrecy capacity observed is almost equal to the perfect secrecy.

Traditional application layer cryptographic mechanisms cannot go beyond detection of signal corruption to determine the eavesdroppers. For the detection of malicious behaviour by the compromised relay node in the cooperative communication, Mao and Wu [24] proposed a cross-layer technique which makes use of adaptive signal detection at the physical layer with the statistical signal detection scheme making use of pseudorandom tracing symbols at the application layer. In [25], two different cooperative techniques are introduced viz. cooperative relaying and cooperative jamming for gaining the security. Power allocation of relay with cooperative relaying and jamming is studied for the achievement of optimum secrecy rate with the available source transmit power. Secrecy capacities of cooperative relaying and cooperative jamming techniques are compared with and without eavesdropper's channel state information.

Imperfect channel condition is the challenging issue for consideration of physical layer security in wireless communications. Cooperative decode and forward approach is considered to combat channel fading and achieve security in the information transmission in [26]. For the assumption of the presence of one eavesdropper, optimal solution is achieved with the help of iterative solution for transmit power minimization consideration. For the assumption of multiple eavesdroppers, due to the problem of secrecy capacity maximization with transmit power minimization, suboptimal solution is proposed by considering the restriction of complete nulling of signals at all the eavesdroppers.

Synergy MAC protocol for now a days Wi-Fi security framework is nothing but the extension of cooperative communication protocol at physical layer to the MAC sublayer. It results into the advantage of spatial diversity with increased transmission rates. For security adjustment in the cooperative scenario, two new security schemes are proposed for 802.11i viz. WPA and WPA2. Various security algorithms such as WEP, WPA and WPA2 are appropriately analyzed in [27] to function with Synergy MAC. The speciality of Synergy MAC is that it has multi-rate capability for packet transmission.

For establishing secure connection in between source and cell edge destination users in the presence of an eavesdropper, relay placement is observed to be more advantageous. Also when path loss is more severe, relay transmission is found to be beneficial. In the randomize-and-forward (RF) relaying mechanism, different randomization is introduced in each hop which is proved to be better physical layer security solution as compared to the traditional decode-and-forward (DF) relay technique [28]. For achievement of physical layer security, two cooperative relaying schemes are analyzed in [29], namely Decode-and-forward (DF) and Cooperative Jamming (CJ). For cell edge users, relays in between decode the received signal and again encoded and weighted signal is transmitted to the receiver. While the source is transmitting the weighted information signal to the receiver, some of the cooperating relay nodes are transmitting weighted noise signal to misperceive the eavesdropper. Two objectives are taken into consideration viz. maximization of the achievable secrecy rate and minimizations of total transmit power.

Due to open nature of multi-hop cooperative communication networks, they are inherently prone to the security threats such as impersonation attacks and message integrity at the receiving end. In [30], the authors have put forth a prevention based technique for secure relay selection for cooperative wireless communication which includes authentication protocol designed with hash chains and Merkle trees. The proposed security system can enhance the number of messages in the Merkle tree and at the same time it can appropriately select secure relay nodes for cooperative communication with significant improvement in the throughput QoS. Throughput attained by using this technique is observed to be higher than the systems without security provision.

In [31], secure cooperative transmission technique making use of physical layer security which considers the presence of passive eavesdroppers. The channels under consideration are frequency flat and frequency selective channels. By exploiting the local information available at individual nodes, full diversity and prevention against malicious behaviour is achieved by keeping intact the transmitter efficiency. The proposed protocol is named as Anti-Eavesdropping Space Time Network Coding (AE-STNC) which works on the principle of randomizing the signals being received at the eavesdroppers

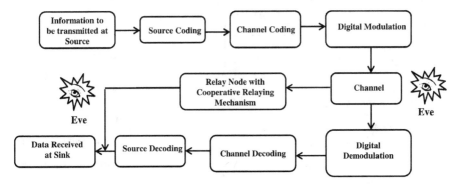

Figure 8 Information theoretically secure cooperative communication link.

with best channel quality so that it becomes difficult for the eavesdropper to capture the messages under transmission. The AE-SNTC protocol is extended further to design AE-STFNC for the broadcast asynchronous cooperative communication networks which is also provides the flexible diversity with security.

Due to highly mobile nature of the mobile adhoc networks, they suffer from imperfect channel conditions and frequently changing topology. The important network design parameters such as security and throughput are simultaneously analyzed in [32] for mobile adhoc networks. The authors have projected a topology control mechanism with authentication for throughput enhancement by combining higher layer security techniques with physical layer security techniques for CWC. The proposed system combines the authentication protocol technique from the upper layers in the protocol stack and transmission methodology from the physical layer to improve the overall cooperative system's throughput.

4 Proposed Secured Cooperative Scheme

Information theoretic source and channel coding techniques can be used with appropriate modulation techniques to achieve perfect secrecy capacities for the CWC. Today green energy is the buzz word and to achieve energy efficiency, we have to think about cost effective security measures. Data compression techniques under source coding mechanism provide high data rates ensuring less energy requirement. For achieving reliability of communication, channel coding methods are extremely useful. Physical layer security

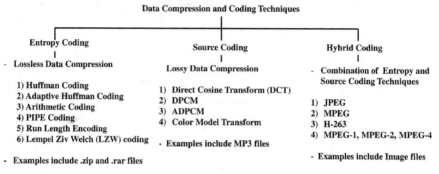

Figure 9 Different data compression and coding techniques.

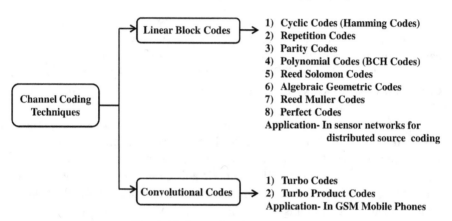

Figure 10 Channel coding mechanisms.

making use of information theoretic techniques ensures security with minimal resource consumption. The proposed mechanism is depicted in Figure 8.

Generally, source coding is done for data compression and channel coding is performed for error detection and correction. For data compression, shortest average description length of a random variable is used. Variable length coding is applied for data compression in which short descriptions are allocated for most frequent outcomes and comparatively longer descriptions are assigned for less frequent outcomes. Data compression and coding techniques include entropy coding, source coding and hybrid coding as shown in Figure 9. It contains different subtypes of these main encoding techniques [33].

In order to achieve reliability of communication in terms of low bit error rate, error control coding or channel coding is the essential technique to detect

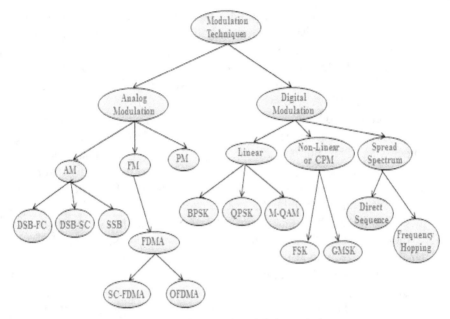

Figure 11   Classification of modulation techniques.

and correct the errors introduced in the channel due to noise [34]. Basic idea behind error correcting codes is nothing but addition of certain amount of redundancy to the message prior to transmission in a known manner. Various channel coding mechanisms are shown in Figure 10.

For conveying a message signal over long distance communication links, some parameters of a low frequency periodic waveform are varied according to high frequency carrier signal attributes [35], which is nothing but modulation. Each of the subclasses of analog modulation technique demand different bandwidth and power for their operation. Analog modulation technique has the disadvantage of more hardware and bandwidth requirement. Once, the noise gets added in the channel, it is carried out as it is till the receiving end since there is no any provision of error control coding for the analog modulation.

Digital modulation has advantages over analog modulation in terms of less hardware requirement, less interference, provision of the error control coding methods and less bandwidth demand. Different modulation techniques are illustrated in Figure 11. Design of a communication system is dependent on the application for which it is to be used. Two major criteria for choosing modulation technique are noise and bandwidth efficiency. Orthogonal frequency division multiplexing (OFDMA) is a good combination of modulation and multiplexing techniques wherein, available spectrum is divided into multiple carriers and each one of them is modulated at a low data rate. Each of these carriers are closely spaced and orthogonal to each other. Hence, OFDMA systems are immune to noise.

Trellis coded modulation (TCM) is a mechanism which combines coding and modulation in a one function. TCM makes use of various concepts of signal processing. Convolutional coding and PSK or QAM modulation schemes are combined in TCM. Ungerboeck's set portioning rules are used for portioning the PSK or QAM sets. TCM is proved to be a bandwidth efficient technique. It promotes highly efficient transmission of information over band limited channels such as telephone lines. It provides almost approximately 40% more spectral efficiency as compared to the popular Reed Solomon channel coding technique [36]. There are various fading channel model assumptions are available in digital communication system design including Nakagami fading, Log-normal shadow fading, Rayleigh fading, Ricean fading and Weibull fading channels. Mainly two channel assumptions are made while designing a communication system viz. Rayleigh and Ricean Fading channels. For line of sight (LOS) scenario, Ricean channel is a good choice and for long distance or non-line of sight case, Rayleigh fading channel is the better fading channel assumption.

5 Information Theoretic Security Measures

1. Entropy of the source $H(X)$ – Entropy is a measure of the uncertainty of a random variable.

$$H(X) = -\sum_{i=1}^{n} p(x_i) \log p(x_i) \tag{1}$$

2. Entropy at the receiver $H(Y)$ – average information per character at the destination.

$$H(Y) = -\sum_{j=1}^{m} p(y_j) \log p(y_j) \tag{2}$$

3. Entropy of the total communication system as a whole $H(X, Y)$ – average information per pairs of transmitted and received characters.

$$H(X, Y) = -\sum_{i=1}^{n}\sum_{j=1}^{m} p(x_i, y_j) \log p(x_i, y_j) \qquad (3)$$

4. Conditional Entropy $H(Y|X)$ – a specific character x_k being transmitted and one of the permissible y_j may be. $H(Y|X)$ gives an indication of the noise (errors) in the channel.

$$H(X/Y) = H(X, Y) - H(X) \qquad (4)$$

5. Conditional Entropy $H(X/Y)$ – a specific character y_j being received; this may be a result of transmission of one of the x_k with a given probability (a measure of information about the source, where it is known what was received). $H(X/Y)$ gives a measure of equivocation (how well one can recover the input content from the output).

$$H(Y/X) = H(X, Y) - H(Y). \qquad (5)$$

The mutual information for main channel is given by

$$I(X; Y) = H(X) - H(X/Y). \qquad (6)$$

Similarly, mutual information for the Eavesdropper's channel is given by

$$I(X; E) = H(X) - H(X/E) \qquad (7)$$

6. The maximum amount of mutual information is nothing but the secrecy capacity for that particular channel.

$$C_{SM} = \max[I(X; Y)] \qquad (8)$$

For security in cooperative communication, we should be able to prove that $H(E/X) > H(Y/X) > H(R/X)$ and $C_{SE} < C_{SM} < C_{SR}$, where C_{SE} is the Secrecy capacity of the Eavesdropper's channel, C_{SM} is the Secrecy capacity of the main or Direct (Main) channel, and C_{SR} is the Secrecy capacity of the Relay channel.

The maximum amount of eavesdropper's equivocation (uncertainty of eavesdropper about the source message) indicates the system security. According to Ciszar and Korner [37], the special case in which the eavesdropper is less capable, that is

$$I(X; E) \leq I(X; Y). \qquad (9)$$

244   *V.M. Rohokale et al.*

Then the Secrecy capacity of a communication link can be given by

$$C_s = \max_{P_x}[I(X; Y) - I(X; E)] \tag{10}$$

6 Conclusions and Future Scope

This research article puts forth a ready reference for the researchers working on cooperative wireless communication and physical layer security. Virtual MIMO or cooperative wireless communication with information theoretic security aspects can prove a cost effective physical layer security solution for today's mobile computing applications. Depending on the application, source coding, channel coding and modulation techniques can be selected which can give satisfactory data rates with ensured secure communication. This work can be extended further for various applications and security measures. Instead of applying channel coding and modulation blocks separately, one can combine both the techniques with a single Trellis Coded Modulation (TCM) block for the added advantage of bandwidth efficiency with higher data rates. Different eavesdropper locations and assumptions may result into better information theoretic security results. Selective relaying mechanism can be considered with relay weights and polarizations for getting transmit power efficiency.

References

[1] P. Liu, Z. Tao, Z. Lin, E. Erkip, and S. Panwar. Cooperative wireless communications: A cross layer approach. IEEE Wireless Communications, 13(4):84–92, August 2006.
[2] H. Katiyar, A. Rastogi, and R. Agarwal. Cooperative communication: A review. IETE Tech. Review, 28:409–417, 2011.
[3] E.C. van der Meulen. Three-terminal communication channels. Advances in Applied Probability, 3:120–154, 1971.
[4] T.M. Cover and A.A.E. Gamal. Capacity theorems for the relay channel. IEEE Transactions on Information Theory, 25(5), 1979.
[5] A. Sendonaris, E. Erkip, and B. Aazhang. Increasing uplink capacity via user cooperation diversity. In Proceedings of IEEE International Symposium on Information Theory (ISIT), August, p. 156, 1998.
[6] Misha Dohler and Yonghui Li. Cooperative Communications: Hardware, Channel and PHY. John Wiley and Sons, 2010.
[7] A. Saleh, A. Rustako, and R. Roman. Distributed antennas for indoor radio communications. IEEE Trans. Commun., 35(12):1245–1251, December 1987.
[8] Xiaofeng Tao, Xiaodong Xu, and Qimei Cui. An overview of cooperative communications. IEEE Communications Magazine, 50(6):65–71, June 2012.

[9] Li Qian, R.Q. Hu, Qian Yi, and Wu Geng. Cooperative wireless communications for wireless networks: Techniques and applications in LTE-advanced systems. IEEE Wireless Communications, 19(2):22–29, April 2012.

[10] Hui Ming Wang and Xiang Gen Xia. Asynchronous cooperative communication systems: A survey on signal Designs. Science China Information Sciences, 54(8):1547–1561, August 2011.

[11] Chen Yongrui, Yang Yang, and Yi Weidong. A cross layer strategy for cooperative diversity in wireless sensor networks. Journal of Electronics, China, 29(1/2), March 2012.

[12] Vandana Rohoakale and Neeli Prasad. Receiver sensitivity in opportunistic cooperative Internet of Things (IoT). In Proceedings of Second International Conference on Ad Hoc Networks, Victoria, British Columbia, Canada, August 2010.

[13] Pei Liu, Zhifeng Tao, Zinan Lin, Eiza Erkip, and Shivendra Panwa. Cooperative wireless communications: A cross-layer approach. IEEE Wireless Communications, August 2006.

[14] Chao Chen, Baoyu Zheng, Xianjing Zhao, and Zhenya Yan. A novel weighted cooperative routing algorithm based on distributed relay selection. In Proceedings 2nd International Symposium on Wireless Pervasive Computing (ISWPC'07), 2007.

[15] Lun Dong, Athina P. Petropulu, and H. Vincent Poor. Cross-layer cooperative beamforming for wireless networks. In Proceedings of Cooperative Communication for Improved Wireless Network Transmission-IGI Global, 2010.

[16] K.J. Ray Liu, Ahmed K. Sadek, Weifeng Su, and Anders Kwasinski. Relay channels and protocols. In Cooperative Communication and Networking. Cambridge University Press, 2009.

[17] S. Abdulhadi, M. Jaseemuddin, and A. Anpalagan. A survey of distributed relay selection schemes in cooperative wireless ad hoc networks. Wireless Personal Communications, 63:917–935, 2012.

[18] Miguel Sepulcre, Jens Mittag, Paolo Santi, Hannes Hartenstein, and Javier Gozalvez. Cogestion and awareness control in cooperative vehicular systems. Proceedings of the IEEE, 99(7), July 2011.

[19] Dejun Yang, Xi Fang, and Guoliang Xue. Game theory in cooperative communications. IEEE Wireless Communications, 44–49, April 2012.

[20] C.S.R. Murthy and B.S. Manoj. Adhoc Wireless Networks Architecture and Protocols. Prentice Hall, Princeton, 2004.

[21] Yingbin Liang, H. Vincent Poor, and Shlomo Shamai (Shitz). Information theoretic security. Foundations and Trends in Communications and Information Theory, 5(4–5):355–580, 2009.

[22] Yi-Sheng Shiu, Shin Yu Chang, Hsiao-Chun Wu, Scott C.-H. Huang, and Hsiao-Hwa Chen. Physical layer security in wireless networks: A tutorial. IEEE Wireless Communications, April 2011.

[23] Vandana Rohokale, Neeli Prasad, and Ramjee Prasad. Cooperative jamming for physical layer security in wireless sensor networks. In Proceedings of 15th International Symposium on Wireless Personal Multimedia Communications, Taipei, Taiwan, September 24–27, 2012.

[24] Yinian Mao and Min Wu. Tracing malicious relays in cooperative wireless communication. IEEE Transaction on Information Forensics and Security, 2(2):198–212, June 2007.

[25] Ling Tang, Xiaowen Gong, Jianhui Wu, and Junshan Zhang. Secure wireless communication via cooperative relaying and jamming. IEEE GLOBECOM Workshop on Physical Layer Security, pp. 849–853, December 2011.

[26] Lun Dong, Zhu Han, Athina P. Petropulu, and H. Vincent Poor. Secure wireless communication via cooperation. In Proceedings of Fourty Sixth IEEE Annual Allerton Conference, USA, September 2008.

[27] Santosh Kulkarni and Prathima Agarwal. Safeguarding cooperation in Synergy MAC. In Proceedings of 42nd IEEE Southeastern Symposium on System Theory (SSST), USA, pp. 156–160, March 2010.

[28] Jianhua Mo, Meixia Tao, and Yuan Liu. Relay placement for physical layer security: A secure connection perspective. IEEE Communication Letters, 16(6), June 2012.

[29] Jiangyuan Li, Athina P. Petropulu, and Steven Weber. On cooperative relaying schemes for wireless physical layer security. IEEE Transactions on Signal Processing, 59(10), October 2011.

[30] Ramya Ramamoorthy, F. Richard Yu, Helen Tang, Peter Mason, and Azzedine Boukerche. Joint authentication and quality of service provisioning in cooperative communication networks. Elsevier Journal of Computer Communications, 35:597–607, 2012.

[31] Zhenzhen Gao, Yu-Han Yang, and K.J. Ray Liu. Anti-eavesdropping space-time network coding for cooperative communications. IEEE Transactions on Wireless Communications, 10(11):3898–3908, November 2011.

[32] Guan Quansheng, F.R. Yu, Jiang Shengming, and V.C.M. Leung. A joint design for topology and security in MANETs with cooperative communications. In Proceedings of IEEE International Conference on Communications (ICC), pp. 1–6, June 2011.

[33] Thomas M. Cover and Joy A. Thomas. Elements of Information Theory (second edition). Wiley-Interscience Publication, 2006.

[34] Shu Lin and Daniel J. Costello. Error control coding: Fundamentals & Applications (second edition). Prentice Hall Series in Computer Applications in Electrical Engineering. Prentice Hall, 2010.

[35] D.K. Sharma, A. Mishra, and Rajiv Saxena. Analog and digital modulation techniques: A review. TECHNIA International Journal of Computing Science and Communication Technologies, 3(1), July 2010.

[36] Gottfried Ungerboeck. Trellis coded modulation with redundant signal sets Part I: Introduction. IEEE Communications Magazine, 25(2):5–11, February 1987.

[37] I. Csiszar and J. Korner. Broadcast channels with confidential messages. IEEE Transactions on Information Theory, IT-24(3):339–348, May 1978.

Biographies

Vandana Milind Rohokale received her B.E. degree in Electronics Engineering in 1997 from Pune University, Maharashtra, India. She received her Masters degree in Electronics in 2007 from Shivaji University, Kolhapur, Maharashtra, India. She is presently working as Assistant Professor in Sinhgad Institute of Technology, Lonavala, Maharashtra, India. She is currently pursuing her PhD degree in CTIF, Aalborg University, Denmark. Her research interests include Cooperative Wireless Communications, AdHoc and Cognitive Networks, Physical Layer Security, Information Theoretic security and its Applications.

Neeli Rashmi Prasad, Ph.D., IEEE Senior Member, Director, Center For TeleInfrastructure USA (CTIF-USA), Princeton, USA. She is also, Head of Research and Coordinator of Themantic area Network without Borders, Center for TeleInfrastruktur (CTIF) headoffice, Aalborg University, Aalborg, Denmark.

She is leading IoT Testbed at Easy Life Lab (IoT/M2M and eHealth) and Secure Cognitive radio network testbed at S-Cogito Lab (Network Management, Security, Planning , etc.).

She received her Ph.D. from University of Rome "Tor Vergata", Rome, Italy, in the field of "adaptive security for wireless heterogeneous networks" in 2004 and M.Sc. (Ir.) degree in Electrical Engineering from Delft University of Technology, the Netherlands, in the field of "Indoor Wireless Communications using Slotted ISMA Protocols" in 1997.

She has over 15 years of management and research experience both in industry and academia. She has gained a large and strong experience into the administrative and project coordination of EU-funded and Industrial research projects. She joined Libertel (now Vodafone NL), The Netherlands in 1997. Until May 2001, she worked at Wireless LANs in Wireless Communications and Networking Division of Lucent Technologie, the Netherlands. From June 2001 to July 2003, she was with T-Mobile Netherlands, the Netherlands.

Subsequently, from July 2003 to April 2004, at PCOM:I3, Aalborg, Denmark. She has been involved in a number of EU-funded R&D projects, including FP7 CP Betaas for M2M & Cloud, FP7 IP ISISEMD ICt for Demetia, FP7 IP ASPIRE RFID and Middleware, FP7 IP FUTON Wired-Wireless Convergence, FP6 IP eSENSE WSNs, FP6 NoE CRUISE WSNs, FP6 IP MAGNET and FP6 IP Magnet Beyond Secure Personal Networks/Future Internet as the latest ones. She is currently the project coordinator of the FP7 CIP-PSP LIFE 2.0 and IST IP ASPIRE and was project coordinator of FP6 NoE CRUISE. She was also the leader of EC Cluster for Mesh and Sensor Networks and is Counselor of IEEE Student Branch, Aalborg. Her current research interests are in the area of IoT & M2M, Cloud, identity management, mobility and network management; practical radio resource management; security, privacy and trust. Experience in other fields includes physical layer techniques, policy based management, short-range communications. She has published over 160 publications ranging from top journals, international conferences and chapters in books. She is and has been in the organization and TPC member of several international conferences. She is the co-editor is chief of *Journal for Cyber Security and Mobility* by River Publishers and associate editor of *Social Media and Social Networking* by Springer.

Ramjee Prasad (R) is currently the Director of the Center for TeleInfrastruktur (CTIF) at Aalborg University (AAU), Denmark and Professor, Wireless Information Multimedia Communication Chair. He is the Founding Chairman of the Global ICT Standardisation Forum for India (GISFI: www.gisfi.org) established in 2009. GISFI has the purpose of increasing the collaboration between European, Indian, Japanese, North-American, and other worldwide standardization activities in the area of Information and Communication Technology (ICT) and related application areas. He was the Founding Chairman of the HERMES Partnership – a network of leading independent European research centres established in 1997, of which he is now the Honorary Chair.

Ramjee Prasad is the founding editor-in-chief of the Springer *International Journal on Wireless Personal Communications*. He is a member of the editorial board of several other renowned international journals, including those of River Publishers. He is a member of the Steering, Advisory, and Technical Program committees of many renowned annual international conferences, including Wireless Personal Multimedia Communications Symposium (WPMC) and Wireless VITAE. He is a Fellow of the Institute of Electrical and Electronic Engineers (IEEE), USA, the Institution of Electronics and Telecommunications Engineers (IETE), India, the Institution of Engineering and Technology (IET), UK, and a member of the Netherlands Electronics and Radio Society (NERG) and the Danish Engineering Society (IDA). He is also a Knight ("Ridder") of the Order of Dannebrog (2010), a distinguishment awarded by the Queen of Denmark.

Vulnerabilities and Countermeasures – A Survey on the Cyber Security Issues in the Transmission Subsystem of a Smart Grid

Yi Deng and Sandeep Shukla

Department of Electrical and Computer Engineering, Virginia Tech, Blacksburg, VA 24060, USA; e-mail: {yideng56, shukla}@vt.edu

Abstract

With the increased investment and deployment of embedded computing and communication technologies in the power system – the smart grid vision is shaping up into a reality. The future power grid is a large cyber physical system (CPS) which is vulnerable to cyber security threats. Among the three major subsystems of a power grid – generation, transmission and distribution – this survey focuses on the transmission subsystem because most of the cyberization of the grid has been happening in this subsystem. This is due to the need for distributed measurement, monitoring and control to retain the stability, security, and reliability of power transmission system. Given the geographically dispersed generation facilities, substations, control centers, data concentrators etc., efficient data communication is required, and therefore large scale networking – either proprietary or leased – is happening. The goal of this paper is not to be comprehensive to include all efforts of securing the transmission system from cyber borne threats, but to provide a survey of various vulnerabilities, and countermeasures proposed by various research efforts. One of the focus area in this survey is the Phasor Measurement Units (PMUs) and Wide Area Measurement System (WAMS) technology – mostly due to our familiarity with the issues for this specific technology deployment – rather than any attempt to indicate that this is the most vulnerable technology in the transmission subsystem. Our hope is that this survey will familiarize any uninitiated reader with the issues and provide incentive to un-

Journal of Cyber Security and Mobility, Vol. 1, 251–276.

dertake systematic research programs to thwart cyber attacks on our national power delivery infrastructure.

Keywords: smart grid cyber security, cyber attacks, synchrophasor technology, phasor measurement unit (PMU), wide area measurement system (WAMS), power system monitoring, power system protection, power system control.

1 Introduction

The impact of the national power grid infrastructure is so deeply rooted in the modern society that we often forget about its importance as we do for the air we breathe. However, the present large and complex power system infrastructure is not built under any top-down planning but rather has evolved from the small system that was built in the late 19th Century. The entire system is continuously upgraded upon the development of advanced technologies through each decade [1]. With the advent of the information age in the late 1970s, people increasingly rely on all kinds of electrical equipments, and on ever-increasing demand for more energy, the power systems' capacity had to be revised dramatically over the years. During the recent three to four decades, the only effective way to achieve the goal of satisfying the growth on demand is either by increasing the power generation capacity, the number of power plants, the transmission line capacity or by limiting electricity usage. Apparently, these kinds of solutions are not satisfactory. To move forward, we need a new power system that is capable of allocating our current power resources intelligently and satisfying the growing complexity and demands of electricity in the 21st Century [2,3].

The concept of a smart grid emerged around 2003 but currently the development and deployment of smart grid projects are in progress world-wide [4]. As shown in Figure 1, the smart grid still retains the legacy of traditional power grid infrastructure such as power generator (circle 1), transmission lines (circle 2), substations, distribution lines (circle 3), transformers, and user terminal equipments (circle 4). In addition, the system integrates renewable green energies, such as solar energy, wind energy, biofuel, wave energy, geothermal energy, and hydro energy to substitute the non-renewable resources. The ultimate purpose of smart grid is to bring economy, security, sustainability, and convenience to both utilities and customers.

Expected to be the next generation of power grid, the objectives of building a smart grid is to maintain an efficient, reliable and secure electricity

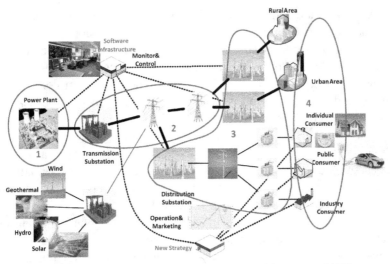

Figure 1 A hierarchical system infrastructure of the smart grid [5].

infrastructure to meet the increasing demand of electricity. The characteristics of a smart grid are defined as follows [6]:

- Utilizing digital, computational, communication and control technologies to create better monitoring, protection and control.
- Integrating the existing hardware and newly developed software to make the system optimal.
- Deployment and integration of distributed renewable resources to reduce the greenhouse gas emissions.
- Employing demand response strategy that makes the electric power dispatching more reasonable.
- Development and deployment of end-user intelligent devices e.g. smart meters to realize the interactive between utilities and consumer devices.
- Integration plug-in hybrid electric vehicles or pure electric vehicles to achieve electricity storage and peak shaving.

Traditionally, the power grid used limited one-way communication so the utility was inefficient to respond to an ever-changing electricity demands. The smart grid is using two-way interaction, where the system information exchanges timely between the power grid operators and the customers [7]. With the help of communication networks, a smart grid control center can monitor and control the real-time parameters of distributed power sources. If upgrading from Supervisory Control and Data Acquisition (SCADA) to

Wide Area Measurement System (WAMS), the smart grid control center is able to acquire the dynamic characteristics of transmission line parameters [8]. Consisting of smart meters, the Advanced Metering Infrastructure (AMI) engages the end users to the smart grid distribution mechanism [9].

However, when the smart grid gets great benefits from computational resources and communication networks, the system will face the risk of cyber attacks and challenges associated with the cyber infrastructure. As many popular website servers may be vulnerable to various types of cyber attacks (such as denial of service attacks), the system control center in a smart grid may become the main object of cyber attacks. Any adversary can attack the system control center by compromising certain numbers of remote sensors that are connected through the same network [10].

Among cyber security issues in the Internet space, one important property is – *information confidentiality*. All the safety measures should be taken to prevent the disclosure of information to unauthorized individuals or systems. Another important requirement – *data integrity* – makes sure that the transmitted data cannot be modified undetectably. A further requirement – *availability* of information ensures that the system must keep the information available when needed. However, in smart gird, the priority of Confidentiality, Integrity, and Availability (CIA) may be considered to be in the reverse order [11]. Among all the key features of a smart grid, the ability to provide a high quality, reliable and sustainable power is the fundamental requirement. The availability of power supply is the most critical quality metric when implementing a smart grid. In a smart grid, the definition of integrity means that the system control center can collect the measurement data accurately, timely, and effectively. Ideally, there should be no transmission error, no false data, and tampered data sent to a system control center. Lastly, the confidentiality in a smart grid determines the privacy issues caused by variety of intelligent devices installed in homes or in substations. The utilities are obliged to protect consumers' personal information, such as telephone numbers, social security numbers, etc., and even prevent from divulging users' personal habits.

At this point it is important to clarify that for the three major subsystems of a power system infrastructure – generation, transmission, and distribution – the cyber security challenges are different. A Generator can be attacked by a breach in the local control room, but can also be destabilized by other means germane to the transmission system vulnerabilities. A distribution substation can also be breached because modern substations are connected to a wide area network to communicate to the control center, and have internal local area network for communication between intelligent electronic devices (IEDs). The Advanced Metering infrastructure (AMI) also communicates through wireless networks to the substations and control stations. The field technicians may also be equipped with network-enabled devices to communicate information on the location of equipments on a distribution network. Thus, various cyber threat models have been identified at the distribution level [12] – but this survey is not intended to cover the distribution system. Our focus is on the transmission system and its security in a smart grid.

A number of threat models have been identified for the transmission subsystem. The malicious data injection attack against the power system state estimation, is well studied in the literature [13]. In this attack, the adversary may execute a joint attack vector on partial meters so that all the bad data detection techniques we are using today will fail to detect the bad data. In [14], the authors defined a special linear injection attack model that mixed with meter measurements, and the bad data detector cannot perceive the false measurement data. By adjusting the attack model, the results of state estimation can be falsified to a certain extent. Meanwhile, there are other kinds of cyber attacks such as denials of service (DoS) attacks, traffic analysis attacks, high-level application attacks, etc., that may affect the security of the power system.

Synchronized phasor measurement technology provides the power system a more precise and real-time measurements for estimating the system state. Global Positioning System (GPS), with the capability of supplying high precision reference clock for other systems, synchronizes the wide area distributed Phasor Measurement Units (PMUs) measurement data. Having the GPS time reference, each PMU can measure the positive sequence phasors of voltages and currents accurately. Even more, PMUs provide the measurements of state vector directly, rather than estimating it from SCADA measurements. By using the PMUs in the smart grid, the operational process for a system control center will change. It increases the accuracy of state estimation and improves the observability of power network. Eventually it can prevent many common cyber attacks. In the next sections we will go into the details of some of these.

This paper is organized as follows. Section 2 discusses the existing cyber attacks against the operation of transmission subsystem of a smart grid. Section 3 introduces the principle of synchrophasor technology and the composition of wide area measurement system, and explores the phasor measurement applications in state estimation, power network observability, and cyber attacks prevention. Section 4 discusses the future directions of using phasor measurement units in the smart grid security research. Section 5 provides concluding remarks.

2 Vulnerabilities in Smart Grid

Through the integration of advanced digital, computational and communication technologies, the centralized power system control center has more ways of managing the entire system. For example, the control center is capable of gathering information from wider geographical area, supporting greater data storage, calculating faster, and sending commands in real-time. Nevertheless, an indisputable fact is that whichever component is critical for system operation, it is weak and easy to become an attacking target. The two-way communication infrastructure provides the adversary with the possibility of attacks.

In the traditional power system, the attackers are difficult to compromise those measurement devices unless they implement a physical damage attack. However, in a smart grid system, the distributed and networked devices provide the interfaces for attacks to access. Wide area communication networks provide a possibility that the attackers can hack into the intra-network by breaking through the intermediate firewalls. Wireless communication networks between smart meters and meter data management center are direct exposure to the attackers. If the attackers can compromise a certain number of remote devices, they can even attack the system control center.

In literature, researchers have found some cyber attacks against power system. There are denial of service attack, malicious data injection attack, traffic analysis attack, and high-level applications attack.

2.1 Denial of Service Attack

The denial of service (DoS) attack is a common attack method in computer-based networks. The DoS attack attempts to prevent the provider from supplying resources and functions available to its users. In communication network areas, the main objects of DoS attack are popular website servers, data centers, wireless communication base stations, etc. The consequences of DoS attack include:

- The computational or communicational recourses of the attack objects are exhausted, and no additional performance to supply the normal services.

- The system configuration information such as package routing information are disrupted, and the information cannot reach to the destinations properly.
- The package state information is tampered, and the system executes a wrong operation.
- Physical damages are applied to the service provider and communication media, so that there are no connections available between the users and provider.

There are dozens of DoS attack methods found in cyber network. Among them, the flooding DoS attack is one of the most common type. The flooding DoS attack blocks the whole network channel by repeatedly sending high-priority data packets to the server, so that the server has no time to respond other requests, such as Internet control message protocol (ICMP) flood and synchronize message (SYN) flood. Furthermore, the properties of the communication infrastructure of a smart gird is real-time, and time-delay sensitive. For some applications, such as adaptive out-of-step,the requirements of end-to-end delays should be less than 50 ms [23]. Therefore, these kinds of applications are more vulnerable to DoS attack.

Take the application of backup relay protection scheme as an example. When a short circuit happens, the protective relays will trip the circuit breaker in order to prevent enlarge the impact. In the current power system, all the relays' trip decisions are determined by their local information separately. In the smart gird, the relays cannot make such hasty decision. A new protection strategy called agent-based backup relay protection scheme shown in Figure 2 improves the relay's reliability. For each relay R_i, it has three protection zones. For instance relay R_1 has zone1 (areas between R_1 and Bus_2), zone2 (areas between R_1 and R_2), and zone3 (areas between R_4 and R_5) respectively. If the errors happen between R_4 and R_5, the executive relays should be R_4 and R_5. In the new scheme, the measurement information from R_1 will also send to the executive relays for helping to reduce false trips by R_4 or R_5. In this case, the DoS attack will affect the agent-based protection scheme seriously. Once the system is paralyzed by a DoS attack, the agent network traffic will be saturated, and it cannot gather enough information from nearby relays. When the command message delay expires, the zone1 relays can only execute the default decisions which are not appropriate in some circumstances.

Defending methods against DoS attacks usually involves using firewalls to detect the attacks, or configuring routers to classify the network channels

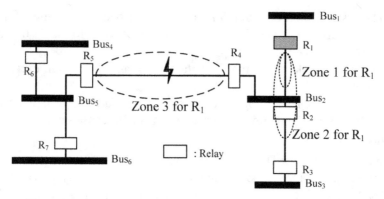

Figure 2 Agent-based supervision of backup relay protection architecture

and block the illegitimate traffic flows. The researchers in [24] did experiments to evaluate the impacts of DoS attack against the transmission delay of communication network in a smart gird.

2.2 Malicious Data Injection Attack

The problem of implementing a malicious data injection attack on power system state estimation was first proposed in [14]. By exploiting the configuration of a power system, an attacker could construct an algebraic attack vector mixed with the compromised meter measurements to introduce state estimation errors arbitrarily without been detected by current bad data detectors.

2.2.1 Power System State Estimation

For a power control center, it is important to monitor the state parameters of the system. The SCADA system in a power grid updates the measurements every 3 to 4 seconds, and yet the system states have actually changed during this period. Therefore, in most cases, the state estimator assumes that the system was in a 'static' state. It uses the active and reactive power flows, bus injections, and voltage magnitudes to calculate the state of a power system. When doing state estimation, the control center assumes that the bad data detector has screened out all the unexpected bad data. Hence the state variables are related to the measurements by the following nonlinear functions:

$$z = h(x) + e, \quad e \sim N(0, W) \tag{1}$$

where $z = (z_1, z_2, \ldots, z_m)^T \in R^m$ denotes the measurement vector acquired from m remote power meters. The z_i are bus voltages, bus real (active) and reactive power injections, and branch real (active) and reactive power flows. $x = (x_1, x_2, \ldots, x_n)^T \in R^n$ denotes the estimated n state vectors. The x_i are bus voltage phase angles and magnitudes. $e = (e_1, e_2, \ldots, e_m)^T$ denotes the measurement error vector introduced by m measurement instruments. The e_i is assumed to be an independent Gaussian distribution with zero mean and a diagonal covariance matrix W. $h(x)$ is a nonlinear function of the state vector x. This nonlinear function is determined by the system parameters and topology of power grid.

In this section, we mainly focus on the DC power flow model, so that the nonlinear state estimation Eq. (1) can be simplified by a linear model:

$$z = Hx + e \qquad (2)$$

where $H = (h_{i,j})_{(m \times n)}$ denotes the measurement Jacobian matrix of size $m \times n$. The Weighted Least Square (WLS) method can solve this DC state estimation problem, which is defined as finding optimal estimate values of $\widehat{x}$ to minimize the target function:

$$J(x) = (z - H\widehat{x})^T W^{-1} (z - H\widehat{x}) \qquad (3)$$

The estimated state vector $\widehat{x}$ is obtained by the matrix solution:

$$\widehat{x} = (H^T W^{-1} H)^{-1} H^T W^{-1} z = Mz \qquad (4)$$

This state estimator is linear, so there will be no iterations. As soon as the measurements are transmitted to the estimator, the estimated state vector can be obtained by matrix multiplication. The Eq.(4) shows that the matrix M is a constant, as long as the system parameters and topologies do not change.

2.2.2 Bad Data Detection

When doing state estimation, the state estimator assumes that all the data candidates are accurate without bad data. Here, the bad data represents the measurements that have problems or errors coming from the measurement units or during the communication process. There are many possibilities to generate bad data. An uncalibrated measuring instrument may cause modest random errors, or a communication error might cause an immense error [15]. Based on the assumption that there is not much connection among these measurement data, that means the measurement data are mutually independent, it is possible to eliminate measurement errors by computing the measurement residuals, which are denoted by $z - H\widehat{x}$.

A common method [16] for detecting bad data is by checking the L_2-norm of measurement residuals:

$$\|z - H\widehat{x}\|_2 = \sqrt{\sum_{i=1}^{m} |z_i - H\widehat{x}_i|^2} \tag{5}$$

or the Largest Normalized Residual (LNR):

$$z_{(i)nor} = [z_i - H\widehat{x}_i]/\sigma_i \tag{6}$$

where σ_i is the measurement variances.

If the received measurements contain bad data, the $L_2 - norm$ or normalized residual of measurements will be abnormal, which is greater than a predefined threshold τ. As usual, the $L_2 - norm$ or normalized residual of measurement residuals follows a chi-squared distribution with $v = m - n$ degrees of freedom [17]. The threshold τ is determined by a hypothesis test with a significance level α. After eliminating the measurement with residuals above the threshold, the estimator will repeat the estimation process without the designated measurements. Then the estimator will do the detection process again, until no bad data was detected.

However, if there are interactions among bad data, the detection performance will dramatically decrease. The bad data can reinforce itself and make the detection procedure to eliminate good data [18].

2.2.3 Malicious Data Injection Attack

Malicious data injection attack was first proposed in [14]. It showed that by compromising enough power meters in a power system, the adversary can manipulate the meter measurements and change the state estimation values arbitrarily without being detected by bad data detection algorithms.

The basic idea of implementing malicious data injection attack is to inject an attack vector $a = (a_1, \ldots, a_m)^T$ into the original measurement vector $z = (z_1, \ldots, z_m)^T$. Let $z_a = z + a$ represents the real measurement data gathered from remote meters. If the state estimator uses the polluted data z_a as the input for state estimation, the estimated state vector should be denoted by $\widehat{x}_{bad}$, otherwise the normal measurement data z should be derived from the normal state vector $\widehat{x}$. The $\widehat{x}_{bad}$ can be represented as $\widehat{x}_{bad} = \widehat{x} + c$, where c denotes the estimation error vector introduced by the malicious data.

With the full knowledge of system parameters H, the adversary can carefully choose the attack vector a. Moreover, if the adversary formalizes the attack model as $a = Hc$, the manipulated measurement data z_a can pass the bad data detection process. Take the L_2-norm detector as an example:

$$\|z_a - H\widehat{x}_{bad}\| = \|z + a - H(\widehat{x} + c)\| \tag{7}$$
$$= \|z - H\widehat{x} + (a - Hc)\|$$
$$= \|z - H\widehat{x}\| \leq \tau$$
$$when \ a = Hc$$

Based on this fact, other researchers did further studies on state estimation related attacks and defenses. In order to find out the minimum number of attackable power meters to make the entire system unobservable, Sandberg et al. [19] defined a security index to characterize the threshold between observable attack and unobservable attack. Regarding the cyber attack problems from the operator's perspective, Bobba et al. [20] found the minimum size set of measurements. If these set of measurements can be well protected, the entire system will prevent from unobservable attacks. In [13] two regimes of attack scenario are considered. One is called – *the strong attack regime* – where the adversary compromises a sufficient number of meters to make system unobservable to the system operator. The other is called – *the weak attack regime* – where the adversary can only control a small number of meters so that the system operator will enlarge its detection capability by using generalized likelihood ratio test (GLRT). On the other hand, the adversary have to trade-off between applying the maximum damages and increasing the probability of being detected by the system operator.In addition, to find the smallest subset of measurements that are well protected to attacks is a high-complexity combinatorial problem and NP-hard, Kim and Poor [21] tried to solve this problem by using a fast greedy algorithm. Giani et al. [22] showed that $p + 1$ secure measurements can neutralize p coordinated attacks.

2.3 Traffic Analysis Attack

In a smart gird system, the information transmitted over communication network should be encrypted. It is difficult for adversary to acquire the contents directly from the raw data. However, the traffic analysis attack is executed by monitoring and intercepting the frequency and timing of transmission messages to deduce the information of the victim networks. By implementation of traffic analysis attack, the adversary can gain the anonymity of some special data packages no matter the packages are encrypted or not. The underlying principle behind the traffic analysis attack is that the analyzed metadata contain the information of sender, receiver, the time, and the length of messages [25]. When the attackers gain the basic network information, they can deduce information about passwords from the interactive between control center and users.

In the application of power system, many system parameters, such as bus voltages magnitude and phasors, active and reactive power are critical for system operation. The distributed monitors will send these system status messages periodically to the system control center. In most of the power systems, the SCADA information will mix with other management data flows into a shared network. By launching traffic analysis attack, the adversary may easily distinguish and isolate these information from other data flows. Then, the adversary may monitor and intercept these critical data, and infer the topology of power grid architecture. In combination with some basic knowledge, such as the serial and shunt admittance of transmission line, the attacker could infer the weakest areas or components in the system, and launch other special attacks to these fatal parts.

For preventing the traffic analysis attack that may analyze the timing and data volume information, a defense mechanism that is based on designing a concatenation of different packets, and implementing random packet drops was proposed [26].

2.4 High-level Application Attacks

The high-level application attacks aim to disrupt not only the basic functions of a power system, such as power flow measurement, state estimation, etc., but also the high-level applications that will execute in Energy Management System (EMS), such as power consumption auto-monitoring, economic dispatch, optimal power flow, adaptive protection, relay protection schemes, electricity real-time pricing, etc. For both consumers and utilities, the economy reason is an important intention of establishing a smart gird. If the smart gird electricity market is under attack, the impacts will destroy the faith of using smart gird.

Take the cyber attack on electricity pricing market as an example. The pricing mechanism in America's electricity market is Locational Marginal Price (LMP) that is usually determining the day-ahead price and the real-time price. In the day-ahead pricing market, the decision principle is matching the supplies and demands, and the LMP is calculated based on the Optimal Power Flow (OPF) results [27]. In the real-time pricing market, the LMP is calculated based on an ex-post formulation [28]. Usually, there are two ways for the adversary to attack the electricity pricing. One direct method is to physically attack or manipulate the electric meter reading to change the quantity of electricity usage. Another indirect way is to compromise the meters in order to affect the LMP calculation. The later attack has broader and more serious consequences.

Researchers have investigated cyber attack issues on power system pricing market. In [29], the malicious data attacks to the real-time electricity market was studied. By attacking the state estimator that can determine the real-time prices, the adversary can influence the revenues of a real-time market. By analyzing an undetectable data injection attack that will manipulate the nodal price of ex-post real-time market, it is shown in [30] that the adversary could gain significant financial profit when the attack is conjunction with virtual biding. As the smart gird is a typical cyber physical system (CPS) with critical infrastructure. The high-level applications attack against any component or application in the system will cause unexpected physical damages.

3 Countermeasures in Start Grid

When facing large number of security issues in smart gird, one possible solution is to learn from the experiences of existing systems that have ever faced the cyber security problems, such as Internet security, IT network security, mobile communication network security, wireless communication security, etc. Most popular network protection technologies, such as firewalls, antivirus, cyber forensics, network identity and authentication, intrusion detection, situational awareness, virtual private network, etc. can be applied in smart gird security design. Moreover, another promising way is to introduce newly developed devices and applications, and construct well defined systems in smart gird to increase the reliability, accuracy, and security of power grid. PMU is such an innovated instrument dedicated for measuring the wide area synchronized phasor to tackle many challenge problems in power system. PMUs based WAMS is regarded as the next generation of wide area monitoring and control system established in smart gird.

3.1 Phasor Measurement Units (PMUs)

3.1.1 Synchrophasor Measurement Principles [15,31]

In power system operation, the coordination of different parts of a power grid is critical. If one of the parts is seriously out of synch or a group of generators going out of step with the rest of power system, the whole system will become unstable and even collapse. Therefore, the power engineers are always eager to monitor the phases of all the bus voltages and line currents in real-time.

In the past, the phasor measurement can only be applied independently. When measuring the phasor of input signal, the phasor measurement unit will sample the analog signal over a finite data window, which is usual one period of the fundamental frequency of the input signal. After digitization, the device applies the Discrete Fourier Transform (DFT) or Fast Fourier Transform (FFT) in practice to calculate the phasor. According to Nyquist criterion, which the sampling frequency should be greater than or equal to twice the maximum signal frequency, there is an antialiasing filter before the data acquisition. The function of antialiasing filter is used to limit the signal bandwidth less than half of sampling frequency.

Since there are N samples of input signal that are taken over one period of the power frequency denoted by $x_k\{k = 0, 1, \ldots, N - 1\}$, the phasor representation is given by

$$X = \frac{\sqrt{2}}{N} \sum_{k=0}^{N-1} x_k \varepsilon^{-jk\frac{2\pi}{N}} \tag{8}$$

Usually, the SCADA system can accomplish the measurement of power flows, but the lack of wide area synchronization mechanism and high-speed communication network make the wide area synchrophasor observation impossible.

The successful use of GPS signal makes the possibility that distributed power system phasor information measures with the same reference time becomes true. As can be seen from Figure 3, the synchronized phasor measurement technology relies on the GPS time signal for supplying synchronous sampling time t' and sequential time stamps.

GPS receivers installed in PMUs provide a precise timing pulse, which keeps the accuracy better than 250 ns and allows the synchronization precisions of local sampling pulse better than one microsecond. Converting to the angle error in a 60 Hz power grid system, the measured phasor errors should be less than 0.02 degrees.

3.1.2 PMU Architecture

A brief block diagram of a generic PMU is shown in Figure 4. The analog inputs are voltages and currents obtained from the secondary windings of the current and voltage transformers. The antialiasing filters block is cooperate

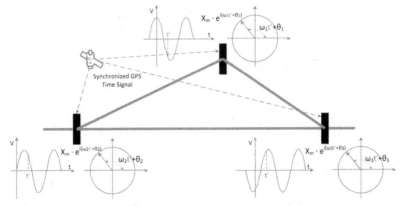

Figure 3 Wide area power parameters sampling with synchronized GPS time signal.

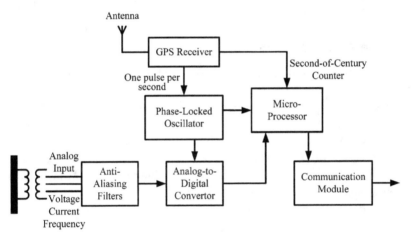

Figure 4 Phasor measurement unit block diagram.

with analog to digital converter (ADC) block to satisfy the Nyquist criterion. In practice, some types of PMUs implement the antialiasing function by using digital antialiasing filter. That means the ADCs first oversample the signal with high sampling frequency and then decimate the signal to a normal rate.

The sampling clock generated from the phase-locked oscillator can be stable. The phasor locked oscillator sync with one pulse per second signal extracted from GPS receiver. Within the modern PMUs, the sampling frequency is 96 or 128 samples per cycle. With the help of digital sampling technology and the higher demand of phasor measurement accuracy, the sampling frequency used in PMUs could be even higher.

After the voltage and current analog signals are digitized, the sampling data will be send to the centralized microprocessor to do the pre-processing. In some cases, the PMU has the ability to store the raw data from ADC to carry out digital fault recorder.

The microprocessor block will first calculate the positive-sequence estimates of all the current and voltage signals. Then assemble the original data with time-stamp. The time-stamp identifies the identity of the universal time coordinated (UTC) clock time, which is used by system control center for sequencing the distributed measurement data. The microprocessor can also process other information such as the local frequency and rate of change of frequency to help local decisions.

Finally, the communication network node is in charge of transfer the time-stamped measurement data compatible with defined IEEE standard for synchrophasors for power system [32].

3.2 Wide-Area Measurement System (WAMS)

The next generation of power system monitoring, protection and control system is wide area measurement system. WAMS is established based on PMUs and other latest communication technologies. Figure 5 shows general hardware architecture of WAMS. The WAMS is composed of five main components: substations with PMUs, substations with PDC, centralized SPDC, relay office, and high-performance regional or wide area networks [33].

Located at the lowest layer of the WAMS hierarchy, PMU installed substations consist of ordinary basic measuring devices including PMUs, digital relays, and intelligent electronic devices (IEDs). Since the volume of transmission data among all the connected equipment is modest, all these devices within substation are connected by the shared media access Ethernet.

In a PDC substation node, there usually is a PDC installed on the substation Ethernet. The first responsibility of a PDC is to gather all the PMU measurement data within the scope of its region. It may send the time-aligned data to the higher level PDC such as a Super PDC, over the network. It may also have the functionality to make certain regional control decisions.

The Super PDC node, which has the capability of storing, analyzing, and illustrating measurement data stays at the top level of the overall architecture. It may be housed in a data center, a phasor data processing center, or a system control center. Most of the monitoring functions, parts of the global protection schemes, and all of the controlling strategies are executed through SPDC.

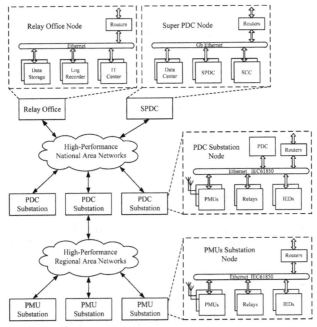

Figure 5 WAMS system architecture.

The communication infrastructure of WAMS can be classified into three types: intra substation local area networks (LAN), high-performance regional networks, and wide area fiber optic networks. The communication standard IEC 61850 defines the mapping of data models to a series of protocols such as manufacturing message specification (MMS), generic object oriented substation events (GOOSE), and sampled measured values (SMV). The high-performance regional networks interconnect several distributed PMUs and one PDC. The highest level of communication network is the most congested network. All the phasor information gathered by PMUs should upload to the centralized system-monitoring center.

3.3 State Estimation with Phasor Measurement

In theory, a PMU installed on one system bus can directly measure the bus voltage phasors and branch line current phasors incident to that bus as shown in Figure 6. Ideally, the wide area measurement system of smart gird is accomplished by deploying PMUs at all system buses. The measured state vector on each bus represents the state of power system at each given in-

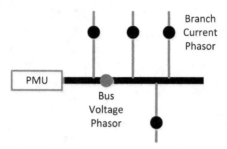

Figure 6 PMU measurements on system bus.

stant. Due to the high updating frequency of measurement data, the dynamic behavior of the power system can be observed directly [34].

The measurement vector consists of synchronized positive sequence voltage and current measurements with zero mean, normally distributed noise component.

$$z_p = \begin{bmatrix} V_p \\ I_p \end{bmatrix} + \begin{bmatrix} e_p \\ e_p \end{bmatrix} \tag{9}$$

The covariance matrix of measurement errors are denoted as W_p

$$W_p = \begin{bmatrix} W_V & 0 \\ 0 & I_p \end{bmatrix} \tag{10}$$

Consider the relationship between voltage measurements V_p and current measurements I_p, the state estimation solution could be solved using weighted least squares method:

$$G \cdot V_p = B^* \cdot W_p + z_p \tag{11}$$

where the G is the gain matrix that is a constant value as long as the system topology is constant. This all-PMU state estimation solution is direct and non-iterative.

In reality, the system does not install enough PMU, the synchrophasor need to mixed with traditional measurements $z_{mix} = [z, z_p]^T$. The hybrid measurement state estimator is presented in [35]. Nevertheless, due to the different nature of complex phasor measurements, the direct inclusion of phasor measurements in state estimators requires significant modifications to the existing EMS software. Zhou et al. [35] used a post-processing algorithm to achieve the estimated states from traditional estimator and then incorporate the phasor measurements.

By integrating phasor measurement data into the process of state estimation, the extra phasor measurement data can improve the network observability so that to prevent from the unobservable malicious data injection attack. Another benefit of using phasor measurement in state estimation is that it improves the bad data detection performance [36]. If the cyber attack is a weak regime attack, the performance of dad data detection scheme determines the attack detection probability. The performance of bad data detection is related to the measurement redundancy, by installing partial PMUs in critical system locations, the bad data detection and identification capability can be improved. So that by utilizing the PMUs especially the secured PMU measurement data can against the dedicated cyber attack on state estimation.

3.4 System Observability with Phasor Measurement

The system observability of a power system means that by installing a certain numbers of PMUs in the power system, all the bus states and branch states can be fully calculated. Baldwin et al. [37] used simulated annealing and graph theory to show that in order to keep the system completely observe, the system have to install PMUs in at least 1/5 to 1/4 of system buses. In an actual large system, the number of PMUs for maintaining the system complete observed is still large. Nuqui and Phadke [38] provided a PMU placement technology, which uses tree search algorithm to optimize the number of PMUs in Depth-of-n incomplete observability occasion.

Using PMU based measurement system provides the possibility of using limited number of secured PMUs to establish a complete observable power network or optimal placing the added PMUs into traditional power infrastructure to make the system from incomplete observability to complete observable.

4 PMU Based Security Issues

4.1 Dynamic State Estimator

Traditionally, the static state estimation is given based on an assumption that the whole system did not change its state during the data scanning interval. Therefore, the static state estimator uses the steady state system model and the SCADA measurements. However, the real practice is that the data scanning takes long enough that the system was actually different. Therefore, when using static state estimation, the system will lose power system details, and have many blind spots when doing false detection. The adversary can attack the system during the scanning interval.

PMU measurements provide a possibility that the system control center is able to tracking the state of system continuously. So that the system control center can monitor and control the electric power system based on the real-time dynamic state. By analyzing the time correlation of measurement data, the dynamic state estimator improves the security performance of anomaly detection.

Dynamic state estimator combines the present or previous state of the power system along with the knowledge of the system's physical model, to predict the state vector for the next time instant [39].When the measurement data acquiring from next instant time arrives, the estimation of system state will be updated to more accurate values. The prediction provides advantages in system operation, power control, decision-making, and attack warning. It sets aside enough time for system control center to take reactions in emergency, and increases the safety sensitivity for any anomaly, such as data injection attack, etc.

The basic dynamic state estimation model is given by [15]:

$$x(k + 1) = x(x) + (\Delta t)r \tag{12}$$

$$z(k) = Hx(k) + v(k) \tag{13}$$

where $x(k)$ is the state at the k th time step; Δt is the time step; r is a maximum rate of change vector, $z(k)$ is the measurement; $v(k)$ is the measurement error. One of methods to solve this problem is using Kalman filtering with the assumption that Δt and $v(k)$ are modeled as zero mean, independent, Gaussian processes.

From another point of view, the dynamic state estimator essentially improves the system's timeframe resolution that prevents the adversary from manipulating the measurement vectors without be observed. There are few researchers exploit the dynamic model of the power network to improve the attack detectability. Pasqualetti et al. [40] showed that for the standard IEEE 14 bus system, it is known that an attack against the measurement data may be undetected by a static state estimator if the attacker compromises as few as four measurements. However, this kind of malicious data injection attack is always detectable by dynamic detection procedure that at least one phasor measurement is measured accurately.

4.2 PMU Data Authentication and Authorization

PMU data authentication and authorization are critical security services for a wide area measurement system, since it enables that the distributed PMUs transfer authenticated measurement data with system control center. Nowadays, the PMUs measurement data is transmitted over the public network, so it is easy for an adversary to manipulate the measurement data without device authentication. An authentication system is a system mechanism where the host service providers may identify their partners in a correct and secure way.

With the assumption that there is no physical layer authentication policy in the system, many cyber attacks that we mentioned above could be successful. If the adversary tampers the measurements with false data, the only method to detect the attack for a system control center is to utilize application layer detection scheme, such as bad data detection. Therefore, establish an authentication scheme is more efficient than post-detection scheme. There are some techniques that could be employed in the data authentication, such as secret password and cryptographic technology, symmetric key based scheme, tokens, etc.

Implementing authorization in PMUs based power system, the system control center prioritizes the distributed PMUs in different levels. The PMUs which are installed in critical places should have higher priority and security than others. They are authorized as first class measurement recourses. In such an authenticated and authorized system, the secured PMUs introduce redundant and trustworthy measurements for defending cyber attacks.

4.3 Spoofing GPS

As we can see from Figure 4 that the PMU GPS receiver provides the one pulse per second for synchronizing the sampling clock, and second of century counter for packaging actual time values into the sampling data. Only through analyzing the data bits of second of century (SOC) and fraction of second (FRACSEC) shown in Eq. (14) [32], the system control center can align all the distributed measurement data. Consequently, the precision of GPS clock time determines the accuracy of PMU measurement results.

$$Time = SOC + Fraction\ of\ Second/TIME\_BASE \qquad (14)$$

It is difficult to acquire and track military GPS signal without encrypted military code (M code) [41], but for the civilian GPS signal, it is vulnerable by inducing a forged GPS signal [42]. For the PMU like civilian GPS receiver, it is hard to detect a spoofing GPS signal because the attacker only needs to tamper the timing information slightly to affect the time accuracy.

The timing information from GPS signals is calculated from two parameters. One is the receiver clock time denoted by T_R that is demodulated from navigation messages with the precision of one second; the other is propagation time denoted by T_P that is acquired from the record of GPS signal propagation with the precision of a millisecond. So that the UTC can be calculated by

$$T_U T C = T_R - T_P - T_C \tag{15}$$

where T_C represents the corrections coming from the GPS receiver.

However, to get the receiver clock time from navigation message, the receiver should first acquire and track the GPS signal by using civilian Coarse Acquisition (C/A) code. To implement acquisition successfully, the receiver needs to search for the code phase of the received C/A code and the Doppler frequency shift [43]. In normal cases, the GPS receiver can acquire the signal by searching the highest correlation peak in the code phase-carrier frequency two-dimensional space [44]. For a GPS spoofer, its task is to mislead the GPS receiver into acquiring a fake signal. If the spoofer generates a new signal that has higher signal to noise ratio (SNR) with higher correlation peak, the GPS receiver will track the fake signal once it lose track caused by intentional signal interference. After that, the timing information calculated from the victim receiver has been manipulated by the spoofer.

Some researchers started to concern themselves with this problems. Gong et al. [44] carried out simulation experiments to assess the impact of time stamp attack to power system transmission line fault detection and location, voltage stability monitoring and location. Humphreys et al. [42] demonstrated a spoofing attack against a GPS time reference receiver installed in a PMU.

Another problem for using PMUs to realize the wide area synchronization is that the GPS signal receiver is the only source for supplying precise time. GPS signal may become unreliable due to weather changes, solar activities, intentional or unintentional jamming, or even worse that the Department of Defense (DoD) changes the GPS accuracy or turns off the civilian signal in some emergency cases. If that happens, the entire power grid system will be be paralyzed, and the security of power operation will be precarious. Therefore, alternative wide area synchronization mechanism should be in consideration.

5 Conclusion

The deployment process of smart gird will enter an explosive growth period during the next decade. From the experience of contemporary Internet, the cyber security issues should be a great deal of attention. In this paper, we introduced the system architecture and characteristics of smart gird. We also discussed the vulnerabilities in smart gird such as malicious data injection attack, denial of service attack, traffic analysis attack, and other high-level application attacks in detail. We introduced the basic concept of synchronized phasor measurement technology and its implementation, reviewed the recent research results that can be used to prevent cyber attacks. Finally, we pointed out the promising research areas based on PMU platform, and discussed potential security issues when establishing WAMS infrastructure. From our observation, we believe that the PMUs based WAMS system applications will play an leading role in the smart gird development, and PMUs based security applications development will become a focal point for smart gird security research.

References

[1] J.D. Glover, M.S. Sarma, and T.J. Overbye. Power System Analysis and Design, fourth edition. Cengage Learning, 2008.

[2] Department of Energy, Office of Electric Transmission and Distribution. "Grid 2030" A national vision of electricity's second 100 years. Meeting Report, 2003.

[3] Department of Energy. The smart grid: An introduction. Report, 2009.

[4] International Energy Agency. Technology roadmap smart grids. Report for International Energy Agency's Energy Technology Policy Division, 2011.

[5] G. Locke and P.D. Gallagher. NIST framework and roadmap for smart grid interoperability standards, Release 1.0. NIST Special Publication 1108, 2010.

[6] United States Congress (H.R. 6, 110th). Energy independence and security act of 2007. [Public Law No: 110-140] Title XIII, Sec. 1301.

[7] Department of Energy. Communication requirements of smart grid technologies. Report, 2010.

[8] A.G. Phadke. The wide world of wide-area measurement. IEEE Power & Energy Magazine, 2008.

[9] D.G. Hart. Using AMI to realize the smart grid. In Proceedings of IEEE Power and Energy Society General Meeting, pp. 1-2, 2008.

[10] S.M. Amin. Securing the electricity grid. The Bridge. Linking Engineering and Society, The Electricity Grid, 40(1):13–20, Spring 2010.

[11] P.D. Ray, R. Harnoor, and M. Hentea. Smart power gird security: A unified risk management approach. In Proceedings of 2010 IEEE International Carnahan Conference on Security Technology (ICCST), pp. 276–285, 2010.

[12] A.A. Cardenas, T. Roosta, and S. Sastry. Rethinking security properties, threat models and the design space in sensor networks: A case study in SCADA systems. Ad Hoc Networks, 7:1434–1447, 2009.

[13] O. Kosut, L. Jia, R.J. Thomas, and L. Tong. Malicious data attacks on the smart grid. IEEE Transactions on Smart Grid, 2(4):645–658, 2011.

[14] Y. Liu, P. Ning, and M.K. Reiter. False data injection attacks against state estimation in electric power grids. In Proceedings of ACM Conference on Computer and Communications Security, pp. 21–32, 2009.

[15] A.G. Phadke and J.S. Thorp. Synchronized Phasor Measurements and Their Applications. Springer, 2008.

[16] A. Monticelli. State Estimation in Electric Power System: A Generalized Approach. Kluwer Academic Publishers, 1999.

[17] A. Wood and B. Wollenberg. Power Generation, Operation, and Control, 2nd ed. John Wiley and Sons, 1996.

[18] A. Monticelli, F.F. Wu, and M. Yen. Multiple bad data identification for state estimation using combinatorial optimization. IEEE PAS-90, 1971.

[19] H. Sandberg, A. Teixeira, and K.H. Johansson. On security indices for state estimators in power networks. In Proceedings of 1st Workshop Secure Control Systems (CPSWEEK), Stockholm, Sweden, 2010.

[20] R.B. Bobba, K.M. Rogers, Q. Wang, H. Khurana, K. Nahrstedt, and T.J. Overbye. Detecting false data injection attacks on dc state estimation. In Proceedings of 1st Workshop Secure Control Systems (CPSWEEK), Stockholm, Sweden, 2010.

[21] T.T. Kim and H.V. Poor. Strategic protection against data injection attacks on power grids. IEEE Transactions on Smart Grid, 2(2):326–333, 2011.

[22] A. Giani, E. Bitar, M. Garcia, M. McQueen, P. Khargonekar, and K. Poolla. Smart grid data integrity attacks: Characterizations and countermeasures. In Proceedings of 2011 IEEE International Conference on Smart Grid Communication (SmartGridComm), 2011.

[23] Y. Deng, H. Lin, A.G. Phadke, S. Shukla, and J.S. Thorp. Communication network modeling and simulation for wide area measurement applications. In Proceedings of 2012 IEEE PES Conference on Innovative Smart Grid Technologies (ISGT 2012), 2012.

[24] Z. Lu, X. Lu, W. Wang, and C. Wang. Review and evaluation of security threats on the communication networks in the smart grid. In Proceedings of the 2010 Military Communications Conference, 2010.

[25] G. Danezis and R. Clayton. Introducing traffic analysis. In Digital Privacy: Theory, Technologies, and Practices, Chapter 5. Auerbach Publications, 2008.

[26] B. Sikar and J.H. Chow. Defending synchrophasor data networks against traffic analysis attacks. IEEE Transactions on Smart Grid, 2(4):819–826, 2011.

[27] E. Litvinov, T. Zheng, G. Rosenwald, and P. Shamsollahi. Marginal loss modeling in LMP calculation. IEEE Transactions on Power Systems, 19(2):880–888, 2004.

[28] T. Zheng and E. Livino. Ex post pricing in the co-optimized energy and reserve market. IEEE Transaction on Power System, 21(4):1528–1538, 2006.

[29] L. Jia, R.J. Thomas, and L. Tong. Malicious data attack on real-time electricity market. In Proceedings of 2011 IEEE International Conference on Acoustics, Speech and Signal Processing (ICASSP), pp. 5952–5955, 2011.

[30] L. Xie, Y. Mo, and B. Sinopoli. Integrity data attacks in power market operations. IEEE Transactions on Smart Grid, 2(4):659–666, 2011.

[31] J. DeLaRee, V. Centeno, J.S. Thorp, and A.G. Phadke. Synchronized phasor measurement applications in power systems. IEEE Transactions on Smart Grid, 1(1):20–27, 2010.

[32] IEEE Power Engineering Society. IEEE standard for synchrophasors for power systems, IEEE Std C37.118TM-2005. 2006.

[33] Y. Deng, H. Lin, A.G. Phadke, S. Shukla, and J.S. Thorp. Networking technologies for wide-area measurement applications. In Smart Grid Communications and Networking, 2012 (to be published).

[34] A.G. Phadke, J.S. Thorp, R.F. Nuqui, and M. Zhou. Recent developments in state estimation with phasor measurements In Proceedings of Power Systems Conference and Exposition, PSCE '09. IEEE/PES, 2009.

[35] M. Zhou, V.A. Centeno, J.S. Thorp, and A.G. Phadke. An alternative for including phasor measurements in state estimators. IEEE Transactions on Power Systems, 21, 2006.

[36] J. Chen and A. Abur. Improved bad data processing via strategic placement of PMUs. In Proceedings of IEEE Power Engineering Society General Meeting, 2005.

[37] T.L. Baldwin, L. Mili, M.B. Boisen, and R. Adapa. Power system observability with minimal phasor measurement placement. IEEE Transactions on Power Systems, 8(2), 1993.

[38] R.F. Nuqui and A.G. Phadke. Phasor measurement unit placement techniques for complete and incomplete observability. IEEE Transactions on Power Delivery, 20(4):2381–2388, 2005.

[39] A. Jain and N.R. Shivakumar. Phasor measurements in dynamic state estimation of power systems. Proceedings of TENCON 2008 IEEE Region 10 Conference, pp. 1–6, 2008.

[40] F. Pasqualetti, F. Dorfler, and F. Bullo. Cyber-physical attacks in power networks: Models, fundamental limitations and monitor design. In Proceedings of IEEE Conference on Decision and Control, Orlando, 2011.

[41] B.C. Barker, J.W. Betz, J.E. Clark, J.T. Correia, J.T. Gillis, S. Lazar, K.A. Rehborn, and J.R. Straton. Overview of the GPS M code signal. In Proceedings of the 2000 National Technical Meeting of The Institute of Navigation, Anaheim, CA, 2000.

[42] T.E. Humphreys, B.M. Ledvina, M.L. Psiaki, B. W. O'Hanlon, and P.M. Kintner, Jr. Assessing the spoofing threat: Development of a portable GPS civilian spoofer. In Proceedings of ION GNSS 2008, 2008.

[43] K. Borre, D.M. Akos, N. Bertelsen, P. Rinder, and S.H. Jensen. A Software-Defined GPS and Galileo Receiver. Birkhauser, Boston, 2007.

[44] S. Gong, Z. Zhang, H. Li, and A.D. Dimitrovski. Time stamp attack in smart grid: Physical mechanism and damage analysis. In CoRR, 2012.

Biography

Yi Deng (M'12) received the B.Eng. and Ph.D. degrees in electrical engineering from Beijing Institute of Technology, Beijing, China, in 2005 and 2010 respectively. He is currently a postdoctoral associate with the

Department of Electrical and Computer Engineering at Virginia Polytechnic and State University in Blacksburg (Virginia Tech). His research interests include synchrophasor measurement technology, power system monitoring protection and control, communication in smart gird, and smart gird cyber security. Dr. Deng's research also covers signal processing, high-performance embedded computing (HPEC), hardware software co-design.

Sandeep K. Shukla (M'99, SM'02) received the bachelor's degree in 1991 from Jadavpur University, Calcutta, and the master's and PhD degrees in computer science in 1995 and 1997, respectively, from the State University of New York at Albany. He is an associate professor of computer engineering at Virginia Polytechnic and State University in Blacksburg (Virginia Tech), where he has been a faculty member since 2002. He is also a founder and director of the Center for Embedded Systems for Critical Applications (CESCA) and director of the FERMAT research lab. He has published more than 150 articles in journals, books, and conference proceedings, and has published eight books. He was awarded the PECASE (Presidential Early Career Award for Scientists and Engineers) award for his research in design automation for embedded systems design, which in particular focuses on system level design languages, formal methods, formal specification languages, probabilistic modeling and model checking, dynamic power management, application of stochastic models and model analysis tools for fault-tolerant nano-scale system design, reliability measurement of fault-tolerant nano-systems, and embedded software engineering. Professor Shukla was elected a College of Engineering Faculty fellow at Virginia Tech in 2004. He is a distinguished visitor of the IEEE Computer Society, a distinguished speaker of the ACM, and a senior member of the IEEE and ACM. He worked at GTE labs and Intel Corporation between 1997 and 2001. He was a researcher at the Center for Embedded Computer Systems at the University of California at Irvine. In 2007, Professor Shukla received a Distinguished Alumni award from the State University of New York at Albany for Excellence in Science and Technology. In 2008, he received the Friedrich Wilhelm Bessel Research Award from the Humboldt Foundation in Germany.

Understanding the Security, Privacy and Trust Challenges of Cloud Computing

Debabrata Nayak

Huawei, Bangalore; e-mail: debu.nayak@huawei.com

Abstract

The overall objective of this paper is to understand the Security, Privacy and Trust Challenges and to advise on policy and other interventions which should be considered in order to ensure that Indian users of cloud environments are offered appropriate protections, and to underpin Indian cloud ecosystem. Cloud computing is increasingly subject to interest from policymakers and regulatory authorities. The Indian regulator needs to develop a pan-Indian 'cloud strategy' that will serve to support growth and jobs and build an innovation advantage for India. However, the concern is that currently a number of challenges and risks with respect to security, privacy and trust exist that may undermine the attainment of these policy objectives. Our approach has been to undertake an analysis of the technological, operational and legal intricacies of cloud computing, taking into consideration the Indian dimension and the interests and objectives of all stakeholders (citizens, individual users, companies, cloud service providers, regulatory bodies and relevant public authorities). This paper represents an evolutionary progression in understanding the implications of cloud computing for security, privacy and trust. Starting from an overview of the challenges identified in the area of cloud, the study builds upon real-life case study implementations of cloud computing for its analysis and subsequent policy considerations. As such, we intend to offer additional value for policymakers beyond a comprehensive understanding of the current theoretical or empirically derived evidence base. which will understand the cloud computing and the associated open questions surrounding some of the important security, privacy and trust issues.

Journal of Cyber Security and Mobility, Vol. 1, 277–288.

Keywords: cloud, security, privacy, trust.

1 Introduction

We identified a number of issues in the literature relating to technological and legal challenges confronting privacy, security and trust posed by cloud computing. Regarding the challenges in the technological underpinnings of cloud computing. There are a number of challenges posed by a range of legal and regulatory frameworks relevant to cloud computing. These include the viability of legal regimes which impose obligations based on the location of data Service models for cloud computing. establishing consent of the data subject; the effectiveness of breach notification rules; the effectiveness of cyber-crime legislation in deterring and sanctioning cyber-crime in the cloud and finally difficulties in determining applicable law and jurisdiction. From an operational perspective, the study uncovered issues relating to the effectiveness of existing risk governance frameworks, whether cloud customers can meet their legal obligations when data or applications are hosted how to be compliant and accountable when incidents occur; whether data will be locked into specific providers; the complexities in performing audit and investigations; how to establish the appropriate level of transparency and finally measuring security of cloud.

Compliance: *Greater harmonization of relevant legal and regulatory frameworks* to be better suited to help provide for a high level of privacy, security and trust in cloud computing environments. For example: *establishing more effective rules for accountability and transparency* contributing to a high level of privacy and security in data protection rules and *expansion of breach notification regimes* to cover cloud computing providers.

Accountability: Improvement of rules enabling cloud users (especially consumers) to *exercise their rights* as well as *improvement of models of Service Level Agreements (SLAs)* as the principle vehicle to provide accountability in meeting security, privacy and trust obligations.

Transparency: Improvement of the way in which levels of security, privacy or trust afforded to cloud customers and end-users can be discerned, measured and managed, including *research into security best practices, automated means for citizens to exercise rights* and *establishment of incident response guidelines*.

Governance: The *European Commission could act as leading customer* by deploying cloud computing solutions as part of its e-Commission initiative and indirectly supporting the improvement of existing operational risk control frameworks. Research funding could be assigned to *improving Security Event and Incident Monitoring* in the cloud amongst other things.

2 Defining Security, Privacy and Trust

Security concerns the confidentiality, availability and integrity of data or information. Security may also include authentication and non-repudiation.

Privacy concerns the expression of or adherence to various legal and nonlegal norms regarding the right to private life. In the European context this is often understood as compliance with European data protection regulations. Although it would be highly complex to map cloud issues onto the full panoply of privacy and personal data protection regulatory architectures, the globally accepted privacy principles give a useful frame: consent, purpose restriction, legitimacy, transparency, data security and data subject participation.

Trust revolves around 'assurance' and confidence that people, data, entities, information or processes will function or behave in expected ways. Trust may be human to human, machine to machine (e.g., handshake protocols negotiated within certain protocols), human to machine (e.g., when a consumer reviews a digital signature advisory notice on a website) or machine to human (e.g. when a system relies on user input and instructions without extensive verification). At a deeper level, trust might be regarded as a consequence of progress towards security or privacy objectives.

3 Risk Control Frameworks

Physical access controls: how can the cloud user achieve requirements for physical access control given the cloud service provider establishes and controls the when who, why and how of physical access measures?

Application development and maintenance: is it possible to assure the development and maintenance of applications in a cloud environment when external parties cloud service provider or other third parties are responsible?

Vulnerability management: assigning responsibility for patch management and the deployment of software and hardware updates between the cloud service provider and cloud user is especially complex given virtualisation and the dynamic reconfiguration of software and infrastructures.

Monitoring: how to establish effective, timely and accurate monitoring of levels of security and privacy in business-critical infrastructure when those responsible for the infrastructure may not be prepared to share such information under standard service level agreement.

Identification and authentication: the integration and control of identity and access management infrastructures in a cloud environment where the cloud service provider might have different approaches and tolerance for risks to identity infrastructure, in addition to the complexities of providing for identity across distributed cloud environments.

Access control: how can the cloud user govern access control risks when the levels and types of access control to key ICT assets deployed by the cloud service provider are unknown.

Encryption – how can the cloud user manage encryption and key infrastructures and assign responsibility across the boundary between their own organisation and the cloud service provider.

Continuity and incident management: how can the cloud user determine appropriate thresholds and criteria for responding to incidents (e.g. agreeing on what constitutes an incident) and policies and processes for responding and achieving assurance of the evidential chain.

4 Solving the Challenges: Observations and Recommendations

Compliance: ensuring that a cloud deployment meets the requirements imposed by the applicable normative framework, including general legislation, sector-specific rules and contractual obligations; the challenges in complying with data protection rules are a key example of this.

Accountability: ensuring that security or privacy breaches in the cloud deployment are correctly addressed, including through appropriate compensation mechanisms towards any victims.

Transparency: ensuring that the operation of the cloud deployment is sufficiently clear to all stakeholders, including service providers and users, both professional businesses and private consumers; this can be witnessed, for example, in the difficulty of determining who/where a cloud service provider is, and where his responsibilities/liabilities end.

Governance: ensuring that the European Commission's policy objectives and actions of the European Commission are well aligned with ongoing stakeholder activities, including by actively participating in the establishment and promotion of standards and best practices, and in interactions with cloud service providers.

5 Growing Focus on Security, Privacy and Trust Concerns in Cloud Computing

Given that the need for public policy depends on the way cloud computing is designed, deployed and used, what is the optimal combination of cloud service provision and governance, taking into account the existing legal and market contexts and the costs, complexity and uncertainties associated with these issues.

Taking into account the incentives of different stakeholders, what is required to ensure that this optimal arrangement can be achieved without adversely distorting the impacts of cloud computing.

6 Identifying Key Issues and Possible Enablers for Security, Trust and Privacy in the Cloud

Assurance of the hypervisors' ability to isolate and establish trust for guest or hosted virtual machines is critical, as this forms the root node for multitenant machine computing and thus could prove to be a single point of failure, since the hypervisor can potentially modify or intercept all guest OS processing.

The same properties of the hypervisor, which enable it to inspect and monitor all processing within and between guest OSs, give the potential for enhanced security monitoring, but will require that current security controls based on dedicated appliances can be migrated to virtual machine architectures. They could also lead to a potential loss of individual customer privacy and security.

For economic purposes, the ability of large-scale instances of virtual machines to be dynamically moved and re-provisioned is vital. It is unclear at this point how adequate the lifecycle management of those instances between hardware and across clouds is, and whether trust can be established to an adequate level, if at all.

7 Security, Privacy and Trust Challenges Inherent to the Legal and Regulatory Aspects of Cloud Computing

The technologically orientated challenges introduced in the previous chapters, it is clear that there are also substantial legal aspects to be taken into consideration for the provisions of cloud computing services. While these challenges are global in nature, the normative response may vary substantially from region to region or even from service to service. Diverging interpretations and legal uncertainties could well endanger the development of innovative cloud service models, as they can adversely affect the trustworthiness of such services: how can users invest in the cloud without a clear perspective on the compliance of the chosen solution with the applicable legal framework, or on the guarantees offered by the service provider.

8 Regulatory Frameworks

In the regulatory framework we should consider the following factors in to account.

- In what country is the cloud provider located
- Is the cloud provider's infrastructure located in the same country or in different countries
- Will the cloud provider use other companies whose infrastructure is located outside that of the cloud provider
- Where will the data be physically located
- Will jurisdiction over the contract terms and over the data be divided
- Will any of the cloud provider's services be subcontracted out

8.1 Regulatory Issues to Be Considered for Cloud

- Indian government Regulation of Investigatory Powers Act
- Stored Communications Act of Indian government
- National Security Letters for investigation
- HIPPA (health-related information)
- GLB (financial services industry)
- state privacy laws
- Video rental records
- Fair Credit Reporting Act

8.2 Establishing the Legal Foundation of Trust: How to Determine Applicable Law in the Cloud

Applicability of the law remains linked to the geographical location of the information society service provider, and in a cloud model it may be difficult to identify this entity or its geographical location. Finally, certain issues, including contractual consumer protection clauses and intellectual property protection, are to be handled very carefully by the regulators.

8.3 Handling Disputes in the Cloud: How to Reinforce Trust by Building in a Mechanism for Accountability

Regulation is linked to the physical location of the stakeholders (typically the place of establishment of the defendant), and certain areas of law are excluded from its scope. Thus, here too, alternative mechanisms of deciding the competent jurisdiction (principally voluntary choice by the parties) will need to be considered, as well as alternative conflict resolution mechanisms, including mediation and binding or non-binding arbitration.

8.4 Cloud Offers the Same Protection of Intellectual Property Rights and Provision of Confidentiality and Data Portability

Software and Database Directives, the IP Rights Enforcement Directive, and several parts of the aforementioned eCommerce Directive, but it is also worth noting that these regulatory measures primarily address intellectual property rights, with rules relating to know-how, trade secrets or confidentiality still being determined largely at the national level.

8.5 Meeting Security Obligations and Responding to Cybercrime

In order for end users to trust cloud services, they must be secure, which implies robustness, reliability and availability. Cloud service providers will need to offer the required guarantees in this regard, by protecting their services against internal threats and against external attacks

9 Cloud Security Advantages

- Exposure of internal sensitive data reduced by shifting public data to a external cloud.
- Cloud homogeneity simplifies security auditing/testing.
- Clouds enable automated security management both internally and externally.
- Redundancy/Disaster Recovery.
- Reduces in-house IT security administration.

9.1 Cloud Security Challenges to Be Taken Care of by the Regulators

- *Trust*

- Putting too much trust to vendor's security model
- *Auditing and investigation*
    - Customer may be out of loop in audit events and findings
    - Obtaining support for investigations at mercy of the provider
    - Logging Challenges
- *Administration*
    - Indirect security administrator accountability
    - Security configurations
    - Identity management
- *Implementation*
    - Black box implementations can't be examined
    - Public cloud vs internal cloud security
- *Data*
    - Regulatory differences and difficulties across national boundaries
    - Data retention issues
    - Data protection in storage and transit
    - Ownership

9.2 Regulator Should Take Care of the Following Points When Locking down the Cloud

- *Securing the cloud*
    - trust
    - multi-tenancy
    - encryption
    - compliance
- *Achieving goals*
    - privacy
    - secure access
    - transparency
- *Trust*
    - Platform trust and trusted computing
    - identity management, user provisioning and access control
    - Federation, control of privileges, SSO
    - Authentication, authorization and auditing

- – Privileged user management
- – Web access management

- *Encryption*

  - – Key management and provisioning
  - – Data leak protection
  - – Data storage and transit Security profile per network

- *Multi-tenancy*

  - – Multi-tenant logging management
  - – Network, VM, Application, process, and data isolation
  - – Security, OS, and Resource Management
  - – Security DMZ per virtual application
  - – Security profile per compute profile

- *Compliance*

  - – Auditing
  - – Log management
  - – Regional/national/international compliances and certification
  - – Legal intercept
  - – Data Privacy

9.3 Compliance and Certification Aspects to Be Taken Care of by the Regulators

- Security related Cloud-specific group
- ITU Cloud Focus Group
- ETSI cloud security group
- SAS70

  - – Auditing compliance

- TIA942

  - – US Data Center

- ISO 27001

  - – Common Criteria certification and compliance

- ISO 15489

  - – Records and Information Management

- LEED

- Leadership in Energy and Environmental Design: green data center
- NIST FIPS 140-2
  - Security Requirements for Cryptographic Modules
- ISA's Security Assurance Certification
  - Embedded Device Security Assessment

10 Conclusion

In this paper we have discussed the pertinent of legal and regulatory domains as applied to cloud computing, most notably relating to legal obligations stemming from location of (personal) data in the cloud, accountability, transparency, consent, security and the definition of responsibilities of those using and processing data. Furthermore, we have discussed the regulatory of operational perspectives, some of the regulators are to manage security, privacy and trust challenges arising in cloud computing deployments. Finally, we discussed how to make cloud a viable and effective business.

References

[1] A. Bisong and S.M. Rahman. An overview of the security concerns in enterprise cloud computing. International Journal of Network Security & Its Applications (IJNSA), 3(1):30–45, 2011.
[2] B. Grobauer, T. Walloschek, and E. Stocker. Understanding cloud computing vulnerabilities. Security & Privacy, IEEE, 9(2):50–57, 2011.
[3] W.A. Jansen. Cloud hooks: Security and privacy issues in cloud computing. In Proceedings of 44th Hawaii International International Conference on Systems Science (HICSS-44 2011), Koloa, Kauai, HI, USA, 4–7 January 2011. IEEE Computer Society, Washington, DC, pp. 1–10, 2011.
[4] Debabrata Nayak. Collaborative security. Paper presented at the Bangalore Security Conference, 10 December 2010.
[5] Debabrata Nayak. Cloud security. Paper presented at Wireless Vitae Conference, Chennai, India, 28 February–3 March 2011.
[6] Debabrata Nayak. Key management in cloud security. Paper presented at ITU, Switzerland, 6–7 December 2010.
[7] Debabrata Nayak. Cloud security. Paper presented at China Shenzhen Conference, 10 December 2010.
[8] Debabrata Nayak. Mobile security. Paper presented at ASSOCHAM Security Conference, India, 1 April 2011.
[9] Debabrata Nayak. Private cloud. Paper presented at Korea-Japan-China Security Conference, 8 October 2010.

[10] Debabrata Nayak. Key management in cloud. Paper presented at IEEE Conference COMSNET, Bangalore, 4–8 January 2011.

[11] Debabrata Nayak. Hybrid cloud security management. Paper presented at Security Conference, China, 12 October 2010.

[12] Debabrata Nayak. Hybrid cloud security management. Paper presented at Security conference, India, 4 April 2011.

[13] Debabrata Nayak, D.B. Phatak, V.P. Gulati, and N. Rajendran. Mobile data networks security issues and challenges. Presented paper at the International Conference on Emerging Technology Bhubaneswar, India, 19–21 December, pp. 137–148, 2003.

[14] Debabrata Nayak, D.B. Phatak, V.P. Gulati, and N. Rajendran. Security issues in wireless local area network. In Proceedings of IEEE Canadian Conference on Electrical and Computer Engineering, 2–5 May, pp. 108–111, 2004.

[15] Debabrata Nayak, D.B. Phatak, V.P. Gulati, and N. Rajendran. Security issues in mobile data network. In Proceedings of IEEE Vehicular Technology International Conference 2004-Fall on 'Wireless Technologies for Global Security', Los Angeles, CA, 26–29 September, pp. 45–49, 2004.

[16] Debabrata Nayak, D.B. Phatak, V.P. Gulati, and N. Rajendran. Modeling and evaluation of security architecture for wireless Local area Networks. In Proceedings of International Conference on Advanced Computing and Communication, Ahmedabad, 16–19 December, pp. 281–285, 2004.

[17] Debabrata Nayak and D.B. Phatak. Modelling and performance evaluation of security architecture for wireless local area networks. Transaction on Engineering Computing and Technology, 3, December 2004.

[18] Debabrata Nayak, D.B. Phatak, and V.P. Gulati. Modeling and evaluation of security architecture for wireless local area networks by indexing methods: A novel approach. In Proceedings of the First Information Security, Practice and Experience Conference (ISPEC2005), Lecture Notes in Computer Science, Vol. 3439, pp. 25–35. Springer, Berlin/Heidelberg, 2005.

[19] Debabrata Nayak, D.B. Phatak, and V.P. Gulati. Performance evaluation of security architecture for wireless local area networks by security policy method. In Proceedings of 2005 IEEE Sarnoff Symposium, Princeton, NJ, USA, 18–19 April, pp.37-40, 2005.

[20] Debabrata Nayak, D.B. Phatak, V.P. Gulati, and N. Rajendran. Policy based performance evaluation of security architecture for wireless local area networks. In Proceedings of 6th World Wireless Congress, San Francisco, USA, 24–26 May, pp. 51–57, 2005.

[21] Debabrata Nayak and D.B. Phatak. An adaptive and optimized security policy manager for wireless networks. In Proceedings of 2007 IEEE Asia Modelling Symposium, Phuket, Thailand, 27–30 March, pp. 155–158, 2007.

[22] Debabrata Nayak, D.B. Phatak, and Ashutosh Saxena. Evaluation of security architecture for wireless local area networks by indexed based policy method: A novel approach. International Journal of Network Security, 7(1):1–14, July 2008.

[23] Debabrata Nayak. Cloud security. Paper presented at ASSOCHAM Cyber Security Workshop, India, May 2012.

Biography

Debabrata Nayak has completed his PhD in wireless security from IIT Bombay. He has been working on security domain in last 18 years. He is a Chairman of Global ICT Forum of India SIG, Co-Chairman for Assocham Cyber Law and IT act, Chairman for Huawei Senior security Expert Consultant Group(R&D), Member of CII (Conferederation of Indian Industry member of International Association for Cryptological Research, Motorola information assurance forum for 3 year, WiMax Forum (GWRG Group), LTE Forum (BWA Group), 3GPP SA3 Security, Cloud Security Alliance, IEEE Security and privacy and Cryptology Research Society of India, and Key member of ITU SG17 Security (Cybex – deals with country specific security), China-Japan-Korea Security Committee, and the ITU Cloud Computing Group.

Debabrata Nayak obtained his Masters Degree from NIT Rourkela, specialized in Elliptic Curve Cryptography and Internet security. Covering wide areas such as Security system Performance evaluation, Design of secure cryptographic system, Wireless Security policy design and implementation. He designed Security solution for INFINET (Indian Financial Network for RBI). He has presented 62 papers in international conferences and technical journals. He was an active member of STIG (DoD) and reviewed guideline for Unix STIG and Network STIG. He has worked with Motorola as Senior Security Architect, Reserve Bank of India as Information Security officer, and with Tata Elxsi as Security expert. He has extensively worked on LTE Security and WiMax Security. He was consultant to various financial institutes for implementation of standards such as BS7799 and ISO 17799. He was also involved in Ministry of Communication and IT of India for Secure mCheque project in IDRBT.

Online Manuscript Submission

The link for submission is: www.riverpublishers.com/journal

Authors and reviewers can easily set up an account and log in to submit or review papers.

Submission formats for manuscripts: LaTeX, Word, WordPerfect, RTF, TXT.
Submission formats for figures: EPS, TIFF, GIF, JPEG, PPT and Postscript.

LaTeX

For submission in LaTeX, River Publishers has developed a River stylefile, which can be downloaded from http://riverpublishers.com/river_publishers/authors.php

Guidelines for Manuscripts

Please use the Authors' Guidelines for the preparation of manuscripts, which can be downloaded from http://riverpublishers.com/river_publishers/authors.php

In case of difficulties while submitting or other inquiries, please get in touch with us by clicking CONTACT on the journal's site or sending an e-mail to: info@riverpublishers.com

www.ingramcontent.com/pod-product-compliance
Lightning Source LLC
LaVergne TN
LVHW012331060326
832902LV00011B/1820